Galactic Travel
at Warp Speed
In Imaginary Time

Charles E. Anzalone

Back Cover

The collage on the back cover contains five radiometers and an eye-catching assortment of unfamiliar objects. The picture symbolizes the diversity of all universal mysteries yet to be discovered like the intricate variety of colors seen through a kaleidoscope. Each object in the collage relates to a galaxy, a planet, a star, or an unknown object floating in space with a lock waiting to be broken and an answer contained within. Like a kaleidoscope, all the answers come in a beautiful array of colors and assortments with interlocking pieces that are part of the bigger picture. Any picture that is awe inspiring and represents a unique concept makes the concept worthy of further investigations.

Library of Congress Control Number: 2008904953
ISBN: Hardcover   978-1-4363-4682-5
                 Softcover   978-1-4363-4681-8

To order additional copies of this book, contact:
Xlibris Corporation
1-888-795-4274
www.Xlibris.com
Orders@Xlibris.com
47723

CONTENTS

CHAPTER 5

CHAPTER 6

CHAPTER 7

CHAPTER 8

INTRODUCTION

The Radiometer

The book cover illustrates a radiometer that is not intended to be a spacecraft. A radiometer is a photon collector making it a solar engine. At three to five inches in diameter, its four vanes can revolve with speeds up to 3,500 rpm. The purpose for presenting the radiometer on the cover is to highlight the idea that space travel beyond the speed of light is possible with similar principles.

A photon is an infinitesimally small particle that originates on every sun and travels with other photons in a packet to form a light beam of coherent light. Coherent simply means that all the photons have the same angular phase. Light from an incandescent bulb is not coherent light because of its random phases. Laser beams and all light beams are a combination of particle motion and electromagnetic propagation. They have mass and frequency! Sunlight and laser beams impact with greater force because both are classified as coherent light.

Constructed from blown glass, the radiometer operates with four vanes on frictionless bearings inside a near vacuum. The vanes are metallic, usually aluminum and painted with a black paint or a black composite on one side. When sunlight strikes the shiny aluminum side, the photons bounce off releasing only a fraction of their energy. When sunlight strikes the dark side, more kinetic energy is absorbed, pushing the vane in accordance with Newton's law of action-reaction. The function of the radiometer has also been associated with heat on the atomic level as per the following statement:

> What really happens is that the faster molecules from the warmer side of each vane strike the edges obliquely and impart a higher force than the colder molecules. The movement of the vane, as a result of tangential forces around the edges, is away from the warmer gas and toward the cooler gas with the gas passing round the edge in the opposite direction. Therefore, the movement is actually due to what happens near the edges of the vanes rather than at the faces.

Warp speed is the speed of an object relative to the speed of light. For example, Warp 1 would equal the speed of light, Warp 2 would equal twice the speed of light, and so on and so forth. A glossary of terms and concepts used throughout these chapters are defined in the back of the book.

This book will investigate similar applications using neutrinos that can propel spacecraft throughout the galaxy in imaginary time. It is backed with suppositions, hypothesis, axioms, and theories that have potential to lead to proven theorems—basically a progression of correlated observations that lead to the possibility that mankind is on the verge of discovering a new frontier.

About the Book and Author

Forty years ago I submitted a feasibility report to the military on the topic that involved the usage of lasers for advanced fuse technology. The military wanted a surface-to-air missile that would detect and destroy incoming enemy missiles. In 1968 the technology I submitted was similar to the concept that *Star Trek* was using in the TV series. My report contained advanced physics formulas and mathematical models based upon actual lab experiments. With four years of field experience and five years of educational experience from the University of Illinois, I was able to write a technical dissertation with math and physics skills that have since grown rusty with time. My boss used to say, "The half-life of an engineer is five years." He was right! At that time, I had hoped to write another report on space travel through other dimensions. I can remember drawing pictures of multiple dimensions on paper that has only two dimensions. That was a time when my skills were strong and my ideas were weak. Now that I'm much older the reverse is true. The formulas contained throughout the book are basic elementary equations, but the ideas and the concepts are unique.

About three years later, the newspapers reported on the military's effort to build a laser gun. In the late 1980's the media reported on the Strategic Defense Initiative or Star Wars. During the time of the Gulf War, the media reported on the military's ability to shoot down Scud missiles. I like to think that my efforts in 1968 cascaded into a defensive posture that was beneficial for mankind. Ideas beget more ideas from other individuals. Good concepts like the ones in this book will grow with the right catalyst to benefit mankind.

This book explains how space travel to other solar systems is possible and what to expect while traveling there. All books, magazines, movies and news documents contain a commonality of truth associated with reality. Ninety-nine percent is the story line but the true realities are expressed in bits and pieces. For example, the scriptwriters for Star Wars develop a story with the concept of warp speed, but they don't define warp speed. Anthropologists discover a new dinosaur skeleton, but they don't say why it got so big and how long it lived before it died. Scientist discover straight lines on the Nazca plains in Peru, but they don't speculate on how they got there. They focus on the monkey drawings and other drawings that human beings made. The bottom line is this. *Before we can learn where we are going, we have to know where we came from.* Our past can predict the future. This book will speculate on the past and try to develop some strong theories on the examples mentioned. The truth need not be a stranger.

Charles E. Anzalone
March 30, 2008

Acknowledgements

Credits go to
- Ms. Rosemary Schaefer who provided the difficult illustrations.
- John at the Lockport Street Gallery in Plainfield, Illinois, who gave me permission to film the inspiring collage inside his store.

Chapter 1

SPACE TRAVEL THEORY

In 1963, I was working toward my bachelor's degree in electrical engineering at the University of Illinois in Champaign-Urbana. A course in atomic physics was a mandatory requirement. As a homework assignment, the professor had asked us to derive the Einstein equation $E=mc^2$. I had no idea where to start because I had believed that Einstein pulled the equation out of thin air! The genius must have said, "Let it be so!"

Previously, the professor had explained that there were two different modes of physics that existed; namely, Newtonian and Einstein. The Newton's formulas started with $F=ma$ and then went on to show the force that existed between charges particles or magnetic poles. When it came to calculating the force between gravitational spheres, Newton's equations were nonexistent. Even Einstein's equations could not account for a gravitational field and its effect upon another heavenly body. Consequently, we need different equations to explain the true physics world.

Before we attempt to hypothesize a new form of physics, we need to state that anything that is unknown has to be solved with a hypothesis before a theory or equation can be proven. This requires an open mind on the subject. The goal is to show that space travel by conventional means with conventional formulas cannot be achieved. The scientist at the Jet Propulsion Lavatories, or JPL, can tell you how much fuel is required to get to Mars and how the cost is astronomical. To travel even farther and quicker, we need a new hypothesis. There is no proof that one exists. A hypothesis can be established by noting the characteristics of unusual phenomena that eyewitnesses have reported when observing UFOs, crop circles, or the plains of Nazca. These observations are a clue to another dimension that we haven't experienced.

Time Travel

Many books have been written and many movies have been made with the topic of time travel as a possibility. It's amazing how intuitive the authors can be in expressing a concept as simple as imaginary time or real time. We all know what real time is a "right now" situation. However, imaginary time is difficult for us to perceive, and yet engineers use the imaginary symbol j to express the reactance of a choke or a capacitor. The symbol j as in +jX or -jX represents a reactance or a form of resistance that is not real. In other words, heat is not generated across a reactance as it is in a resistor. The symbol j also has the numeric value $\sqrt{-1}$ or the square root of minus one, which can't be reduced any further. Up until now, no one has used the expressions +jT or -jT with the symbols representing imaginary time into the future or the past respectively.

Advanced calculus teaches us that we live in a world of three dimensions with each dimension being at right angles to one another. If another dimension can be positioned at a right angle to the existing three, it would be considered a fourth dimension, and "time" is the best contender. At this point, a graph is needed to visualize the relationship between time, light, and gravity. We will allow gravity and light to be the dependent variables plotted against the independent variable or imaginary time.

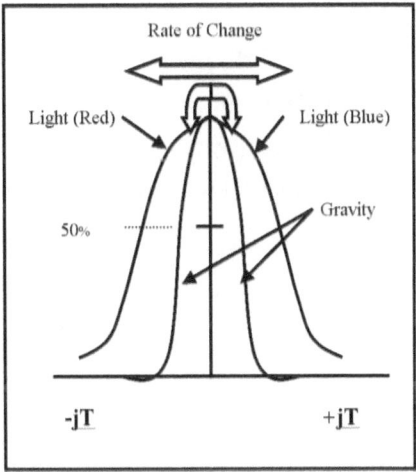

The figure to the right illustrates the reduction of gravity as time progresses from real to imaginary. If a time traveler moves farther to the right or +jT, he/she will experience a visual effect with everything outside his/her craft in a shade of blue. Converesely, traveling in the opposite direction will cause everything to be in a shade of red. Traveling farther in either direction will cause gravity to fall to 50% of its initial value and then to a value of zero.

Zero gravity is achieved by traveling farther along the imaginary time line. When this happens, there is no attraction between a spacecraft and Mother Earth.

Objects on Earth can be seen by the traveler as *Earth moves away* from the traveler at a speed of one ten thousandth the speed of light or approximately, sixty-six thousand miles per hour. If the space traveler wants to hover above Earth, he/she needs to alternate between +jT and -jT. The speed of alteration or the *rate of change* will cause a flickering change of colors across the spectrum between red and blue. Time travelers will experience the color change and so will spectators on Earth observing the craft. At some point along the imaginary time line, radar detection will drop out, but visual contact will still be present. The units for the vertical coordinate on the chart are expressed in percent change with 32 feet/sec/sec representing the initial value for gravity. In addition, the vertical cordinate is normalized to represent the change in color, expressed as delta angstrum. An angstrum is a unit of measure for frequency. The curved lines for gravity and light frequency can be derived from modified exponential and linear expressions, respectively. Gravity is a second-order expression, and light frequency is a first-order expression. Consequently, the light frequency curve will always over shadow the gravity curve as per math expressions. When gravity is at 50% of its initial value, light will have changed by "x" angstroms, where "x" is unknown at this time. In other words, the observer in the craft and the observer on Earth will see each other with a green blue hue in the +jT direction or a orange red hue in the -jT direction. This is what causes UFOs spotted over Mexico City to flickered in different changing colors. By working with the second-order derviation for gravity and the first-order derivative for light in angstroms, a mathematician could approximate the changed value for light in the chart at the 50% gravitational mark. For greater accuracy, emperical measurements would have to be made from within a spacecraft. For any given value on the imaginary line or x-axis, there is a difference between the two curves on the y-axis. This difference tells the space/time traveler where he is, a location marker analogous to bookmarker. This is why crop circles exists and serve as a bookmarker for the space/time traveler when they are preparing for a launch. The change in color of the crop circle provides visual guidance to acertain the amount of reduction in the gravational pull of Earth. This information is critical as the spacecraft begins to accelerate to 0.01% the speed of light.

First, we have to show how it's possible to start traveling at speeds close to the speed of light. The enclosed chart is tabulated here for convience and again in figures 5a and 5b. The chart shows the tangental speed of all the planets. Earth is traveling more than 0.01% the speed of light [or100%*66,243/(186,000*3600) where 186,000 miles per second is the speed of light].

Planet	Distance from Sun (mi*10^6)	One Year (days)	Speed (mi/hr)	Speed (ft/sec)
Mercury	036	088	107055	157000
Venus	067	224	78,594	115647
Earth	092	365	**66,243**	96,916
Mars	141	687	53857	79,371
Jupiter	484	4332	29,229	43,078
Saturn	888	10752	21,631	31,901
Uranus	1784	30663	15,229	22,451
Neptune	2799	60140	12,186	17,961
Pluto	3675	90717	10,606	15,634

When the spacecraft is activated for zero gravity pull, Earth moves away at 66,243 miles per hour. Then, to achieve higher speeds, the spacecraft needs to switch to real time and let the gravitational force of Earth pull it back faster. The gravitational force is now represented by F=ma, where "m" is the mass or weigth of the spacecraft and "a" is Earth's acceleration of 32 feet/second/second or one g-force. A one g-force is the same as twenty-two miles per hour increase for every second that passes. The calculation is as follows:

One g-force = (32 ft/sec/sec * 3,600 sec/hr) / 5,280 ft/mile = 22 miles/hour/sec

Let's assume that the pilot leaves Earth and switches to imaginary time that last for ten hours before switching back to real time. After ten hours, the distance between Earth and spacecraft will be 662,000 miles or beyond the orbit of the moon at 255,000 miles. The pilot switches to real time. The spacecraft begins to accelerate toward Earth and reaches a speed of 662,000 miles per hour in 30,000 seconds or about 8.3 hours. Without any wind resistance, the spacecraft is catapulted toward Earth. The total vector speed is 728,000 miles per hour (or 662,000 + 66,000). Suddenly, it's either a collision like the one in Roswell or signpost up ahead that tells the pilot to do something immediately before the ship crashes. The signpost are the crop circles in the wheat fields. With special optics, the pilot can see that the crop circles are brown as they were when he left them in real time. He avoids the crash by switching to imaginary time and notices that his spacecraft has made the correct transition because the crop circles are now blue due to Doppler effect. After the pilot lays in the coarse, the computer and optical scaner measure the frequency shift of the blue light that corresponds on the chart for zero gravity. The color of the crop circles in the correct shade of blue confirms the moment when the spaceship is ready to travel *through* Earth. At this point, nothing gets in the way because there's nothing there. There aren't any asteroids, comets, or planets, only shadows and the light that they reflect. The Earth can be seen, but it's *not physically there*! Consequently, the spacecraft travels through Earth and out the other side at 728,000 miles per hour or 0.1% the speed of light. Traveling through imaginary time is equivalent to traveling through hyperspace, as coined by writers for the *Star Trek* movie. If Earth and Mars are at their closest, a trip to Mars would take only 67 hours [or (141,000,000-92,000,000)/(728,000)] plus the the initiation time of 10 and 8.3 hours, manuvering time and acceleration time. Landing time is not calculated.

[Author's comment: This is a difficult concept to imagine, but the only circumstantial evidence is superlightning or possibly the events associated with the Bermuda Triangle. Superlightning has been documented and recorded as shooting upward, beyond the cloud layer, thousands of feet. Superlightning shoots away from Earth, following an ion path of a space vehicle as it leaves Earth.]

After reviewing the characteristics of our solar system in the chart above, it becomes obvious that Earth offers the best transportation highway, coupled with visual effects. A combination of tangential speed and gravitational pull of Earth offer the best vector addition for accelerating a spacecraft through space with speeds approaching the velocity of light. An additional benefit are the landmarks that Earth offers, whether they be crop circles, Stonehenge, or even pyramids.

After achieving sublight speeds, the second objective is to attain speeds much greater than light speeds. If travel to Mars at 0.1 % the speed of light takes sixty-seven hours direct travel time, it can be shown in the next chapters that Warp 10 is possible, and direct travel time would take twenty-four seconds [or (67 hrs.*60 min.*60 sec.)/(1,000*10) where Warp 10 is ten thousand times faster than 0.1% the speed of light]. Figure 1a contains calculations for space travel to distant destinations and the associated time involved. The acceleration and deceleration times would take years to accomplish because the human body can withstand only an 8 g-force for short periods of time.

The following chapters discuss the concepts and the methodology required to meet these objectives. Each succeeding chapter will offer a secondary discussion of the most important concepts that were presented in the previous chapters. For example, the chart for gravity versus imaginary time will be revisited because of its significance and the ability to achieve galactic space travel. The chart is so important that it has to be referred to with a name. So let's use GIT to represent the curve as "gravity versus imaginary time."

Chapter 2

SPACE TRAVEL AT
LIGHT SPEEDS AND GREATER

Background Information

The previous chapter developed a scenario with a spacecraft traveling at 0.1% the speed of light as it left the proximity of Earth. Similar to the operation of an auto transmission, a spacecraft would have to shift up to the higher speeds before it could attain light speeds and greater. The value of 0.1% is speculated as a threshold required for achieving greater speeds. Before starting this feasibility study, a review process is in order.

Review

The August 2006 issue of *Discover* magazine has an article titled, "The Einstein Dilemma: Was He Wrong About Gravity?" The article ends with the statement, "Astronomical evidence gathered over the last thirty years suggests that most of the mass in our galaxy is invisible to our telescopes." The operative word in this conclusion is "invisible." We will speculate on what invisible could really mean.

Before starting this dissertation, we have to reflect on Mr. Rod Serling's opening statement in the TV series *Twilight Zone*, "You unlock this door with the key of imagination." With physics, facts, and keen observations, this feasibility study will attempt to unlock that door with imagination. Without creating a parody on words, we have to define "imaginary time" as the time line experienced in the fourth dimension. But first, let's start with Einstein's famous equation; namely, $E = mc^2$.

Theory Explanation

When I was a student at the University of Illinois, a homework assignment in atomic physics led me to believe that I could achieve the correct answer for the equation by simply plugging in the variables for mass and the speed of light. As every engineering student learns, E is not equal to mc². It's equal to delta m times c² or $E = \Delta mc^2$. Delta represents the change in mass. Einstein postulated that the speed of light or c is a universal constant. The following chapters will postulate that the speed of light is constant in specific sectors of the galaxy and changes from sector to sector. The correct equation for energy is written as $\mathbf{E = \Delta m * (c|k)^2}$, where |**k** signifies a constant for a specific sector of the galaxy.

When a log burns, it gives off heat, light, and gasses. It leaves behind ashes. When a hydrogen atom on the sun converts to a plasma and then doubles up with another atom and two neutrons, it changes to a helium atom of lesser weight in a nuclear explosion, producing all the by-products. The hydrogen atom consists of a proton and an electron with an atomic weight of 1.0079 and an atomic number of 1. In the plasma state, the electron is lost, and only the proton with a positive charge remains. The neutron has the same mass as the proton and exists in the sun with a zero charge as per scientific beliefs. The atomic weight of the electron is insignificant.

The latest scientific theory on atomic fusion consist of two protons and two neutrons combining to form a helium atom in the plasma state with an atomic weight that is something less than 4. In other words, $1+1+1+1 = 3.8 +$ nuclear explosion. (The number 3.8 is a number *picked at random* in order to maintain a logical train of thought. Other random numbers in this report will be referred to as a PAR number.)

A newer approach is to imagine a proton taking on a corkscrew trajectory in the process of losing its electron and then being bombarded by a free neutron projectile flying around inside our sun. At this point, either one of two events can happen. The neutron hits the proton, condition 1, or it misses, condition 2. If it misses, it can go on its merry way and nothing happens, or it can assume an orbit around the proton in the same fashion as a bipolar star. Now we can assume that first condition occurs in one out of one thousand cases, where 1,000 is a PAR number. The neutron hits the proton with a violent explosion that causes another chain reaction. In this case, the mass of the neutron and the proton are devastated as much as the burning log in the paragraph above. This has got to be a factual situation on the sun because science tells us that protons convert into photons and neutrons convert into neutrinos. We know that burning logs convert into ashes! After the energy conversion from logs to ashes or protons to photons, the mass of the residue is much smaller. This process has to be more common than the fusion process because the sun produces more energy when the entire atomic mass of the photon is converted. Only a number of photons remain with atomic weight that is insignificant. One proton converts into a number of photons that travel through space in groups called packets or sunlight.

Now, here's the interesting part. Photons and neutrinos have zero rest mass! When the by-products of Einstein's equation, $E = \Delta mc^2$, are extrapolated, it becomes well-known that the mass increases as the speed of light is approached. It also takes a lot of energy to attain the speed of light! The photon and the neutrino attained their energy during the atomic explosion.

We know that a photon is a beam of light, and science tells us that light is coherent and travels in packets of energy. We have to assume that one proton was hit by a neutron and blew up into one hundred photons and one hundred neutrinos, where 100 is another PAR number. The one hundred photons are all traveling together in one bunch or packet with the same phase. To summarize, one photon is a particle traveling in a corkscrew trajectory at the speed of light. Henceforth, light has a frequency. In this report, we'll refer to this as *single* light. It exists in real time. But there is another type of light that we will call *double* light, and it exists in imaginary time.

In condition 2, we talked about the possibility of a neutron assuming an orbit around a proton prior to nuclear explosion on the sun. The probability of the neutron missing the proton and assuming a bipolar orbit is much greater. Hence, the miss-to-bipolar-orbit condition occurs one thousand times more often than condition 1. The next events should be presented in pictorial or animated form. Unfortunately, that can't happen here. Consequently, best verbiage is provided to visualize *double* light traveling in the fourth dimension:

- A proton is traveling in the sun in a corkscrew trajectory.
- A neutron with the same weight as the proton starts a bipolar orbit around the photon.
- A nuclear explosion occurs when second neutron hits first neutron or proton.
- Proton and neutron are converted to one hundred pairs of photon and neutrino respectively and accelerate to light speed.
- Photon has zero rest mass but retains a slight positive charge, retained from the proton parent.
- Photon travels away from the sun creating a magnetic field analogous to a charged particle traveling through a coil. Three dimensions are required to sustain this magnetic field. (To picture a field in three dimensions, one has to visualize a watermelon traveling through space along x-axis with magnetic lines of flux shooting out over the surface along the y- and x-axis.)
- The orbital path of the neutrino satisfies the need for a fourth axis as it circles to proton. (To picture this orbit, one has to picture a cowboy on his horse, riding out of the shoot with a lariat circling over his head. The photon is to cowboy as the neutrino is to the lariat.)
- This visual concept will now be referred to as *double* light.

- Since *double* light is traveling in the fourth dimension, it can't be seen be anyone in real time.

To prove the existence of *double* light, we would have go to imaginary time and witness it. In chapter 1, a chart was presented for gravity versus imaginary time and its relationship for light versus imaginary time. There is a point on the chart where gravity fades to a zero value and light continues to have some value. This is the area defined as *double* light.

Justification

The justifications for *double* light and the fourth dimension are as follows:

1. My professor in advanced calculus class said, "Three dimensions in space are at right angles to each other. If anyone can show where a fourth axis could be at right angles to the other three, you could have a fourth dimension." The cowboy is moving in three dimensions, and the lariat is at right angles to him in a fourth dimension.
2. My other professor in fields and waves class said, "Electromagnetic propagation is a process analogous to a chain with multiple links where one link is an H-field or magnetic field, and the next link is at a right angle to the first, and it represents the E-field or electrostatic field."
3. Mr. Stephen Hawking discusses in his book,[1] with scientific knowledge, that a packet of light can travel through two-slit openings creating an interference-enhancement waveforms and that a single photon can do the same. How can one particle go through

---

[1.] Stephen Hawking, *A Briefer History of Time*, (New York: Random House, 2005).

two openings? Answer: The existence of *double* light has to be the answer.

These justifications lend themselves to discussions on invisibility, dark matter, dark energy, imaginary time, galactic sector time, and most important "traveling at speeds greater than light."

Photon Detectors or Radiometers

Photon detectors do exist. They can be seen inside a three-inch quartz sphere, where 3 is a PAR number. They consist of four diamond-shaped vanes made from aluminum with black paint on the backside. The vanes are mounted on frictionless bearings and they spin around rapidly in the vacuum of the sphere.

Neutrino detectors do exist. At last count, there were three throughout the world. It is said that millions of neutrino travel through one's fingernail in a flash instant. Neutrinos are much harder to detect because of the properties that they exhibit. And here is why: "The majority of the neutrinos are traveling at speeds much greater than the speed of light."

Einstein said that nothing could exceed the speed of light. If we're talking about real time, he was right. If we're talking about imaginary time, he was wrong. Take the case of the cowboy coming out of the shoot with the lariat circling over his head. If the cow and cowboy are traveling at the speed of light, will he be able to lasso the cow? The answer is unknown if the situation exists in real time, but the answer is a definite yes if we're talking about imaginary time. Here's why. When a neutrino is traveling the same elliptical path as the lariat around the cowboy, it will have different speeds with respect to the forward direction. When the neutrino is in front of the cowboy or directly in back, the forward speed of the neutrino will be the same as the speed of the cowboy as he travels the speed of light. That's a given. When the neutrino is accelerating alongside the cowboy, its tangential speed is much greater. So how much greater is this slingshot effect? At this point, we need a scientist with research funds to calculate the speed of an electron in a hydrogen atom and then transpose that to the speed of a neutrino released from its bipolar orbit. If I had to create another PAR number, I would use100 and say that a neutrino travels at one hundred times the speed of light in imaginary time. We have to assume that photon and neutrino are separated by some distance just as cowboy and lariat would be separated. Statistical results can be demonstrated with a histogram to show the release point occurs when the neutrino is trying to "catch up." Some undocumented reports state the explosion on the sun propels particles at seven hundred thousand miles per second or four times the speed of light. When the vector speed, angular speed, and explosive speed of the neutrino are summed up, speeds one hundred times the light speed are feasible.

Chapter 4 discusses histograms and the distribution of neutrinos into different speed categories as per a natural process. The percentage of the neutrino population traveling at the same speed as light is miniscule. Very few travel at this speed (Warp 1) and make the transition to real time where they are detected. This is why the detection of neutrinos is rare. The vast majorities pass through Earth at speeds greater than the speed of light and go undetected. Where do they go? The migration of neutrinos throughout the galaxy is from each sun toward the black hole in the center of the Milky Way. Neutrinos traveling in imaginary time with speeds beyond Warp 1 increase in mass and are pulled toward the black hole. Neutrinos that exit all suns on the side closest to the black hole have a shorter path to travel than the ones that exit on the far (or distant) side. The neutrinos that exit on the far side assume a path with a curvature that may be measurable. If the curvature starts at the same orbital distance that Mercury is from the sun, then more neutrinos will be counted when Earth is positioned between the sun and the black hole than the position that occurs six months later.

[Author's comment: The author can envision streams of neutrinos emanating in a radial pattern from every star in the galaxy. At some locations in the galaxy, they cross over each other's path like the crosswinds on Earth. Some of these streams bounce off walls or barriers in different sectors and create faster streams or what scientists refer to as wormholes. These names such as walls, barriers, wormholes, or aisles all have the same connotation. They encapsulate each spiral sector of the galaxy and will be discussed again in chapter 6.]

Conclusion

We have shown that existence of *double* light is much greater than *single* light because the probability of a neutron colliding with a proton is more difficult than achieving a bipolar orbit. Consequently, the existence of neutrinos in imaginary time has to be overly abundant. Traveling through space at speeds greater than light needs to be accomplished by sailing. Jet-rocket power is not needed. A sail with a specific material that detects neutrinos will do the job. Science tells us that neutrinos can travel through lead, Earth, etc. If neutrinos can't be stopped, then we can conclude that they travel throughout a universe that is frictionless! This conclusion infers that the only particles that exist in imaginary time are neutrinos and photons or light beams.

Achieving Warp Drive Speeds Greater than the Speed of Light

The previous chapter developed a scenario with a spacecraft traveling at 0.1% the speed of light. The surface of the spacecraft would have to be designed so that it has the characteristics of a foil embedded with a relief grate (or valve) to regulate the passage of neutrinos. A spacecraft with the proper element materials traveling at 0.1% the speed of light should be able to stop more neutrinos than our Earth does, traveling at 0.01% the speed of light. At these speeds, the neutrino won't go through the sail but instead push it.

If we assume that a traveler can withstand 8 g's of acceleration, let's calculate how long it would take to accelerate to the warp speed of light or 186,000 miles per second. Calculations in the last chapter showed that 0.1% the speed of light was equivalent to 728,000 miles per hour. The time duration for achieving light speed can be calculated as follows:

$$t = (v2 - v1)/a \quad \text{where} \quad v2 = 186,000 \text{ mi/sec or } 669,600,000 \text{ mi/hr}$$
$$v1 = 728,000 \text{ mi/hr and}$$
$$a = 8 \text{ g's}$$
$$= 8 * 22 \text{ mi/hr/sec}$$
$$= 176 \text{ mi/hr/sec}$$
$$t = (669,600,000 - 728,000)/176 \text{ seconds}$$
$$t = 44 \text{ days}$$

If the human body can withstand an 8 g-force for prolonged periods, it would take 44 days to reach a speed of Warp 1 and 440 days to reach Warp 10.

[Author's comment: I find this 40- or 44-day concept interesting because of a corollary to the Bible, Genesis 7:12, "And the rain was upon the earth forty days and forty nights." The Bible refers to two-time periods of confinement in a ship; namely, 40 days and nights of rain and then 40 days of flooding.]

Summary

In the first chapter, we illustrated a chart with a GIT curve that demonstrates how gravity diminishes to zero as imaginary time is shifted in the positive or negative direction. This area has been referred to as imaginary time where Einstein and Newton laws of physics do not apply. Our calculations for acceleration beyond the speed of light may require a multiplication factor that is not linear. In the real world, engineers can calculate bandwidth shrinkage factor when a RF signal passes through multiple number of filters with the same bandwidth. They can calculate the transmission of an RF signal through a coax cable and show how its speed is 60% slower than traveling through free air. So who is to say that linear relationships for acceleration when traveling in imaginary time can be justified. They can't! Specifically, accelerating to the speed of light in 44 days in imaginary time may be accurate, but accelerating to one hundred times the speed of light in 4,400 days may not be a linear relation as per Newton's laws.

[Author's comment: In other words, the effects of acceleration may be diminished be some shrinkage factor so that travel to more distant solar systems is feasible.]

We have justified how a spacecraft can "sail" at the same speed as neutrinos or one hundred times the speed of light. The speed can be regulated the same way as a sailboat. The real problem is that the human body cannot sustain 8 g's for 4,400 days to meet that goal! If we assume that warp speeds are linear, then we might have to limit the top speed that the human body can withstand to Warp 10.

What would it be like?

Imagine the characteristics of traveling through space in imaginary time at multiple times the speed of light.

1. If imaginary time were a frictionless world, there would be no need to worry about crashing into a planet, asteroid, or sun. We would sail right through them.

2. The problems associated with radiation poisoning wouldn't exist because the only two particles that would exist are protons and neutrinos that are so much smaller than gamma rays and X-rays.

3. Looking out the portal, we would see *double* light, which looks like ordinary light coming from every star. *Single* light would bleed into imaginary time giving everything a slight color cast.

4. Critical navigation would be required to land on a planet in another solar system. The sail has to be trimmed, and a breaking action has to be applied by jockeying in and out of real time. Any slight error could result in an explosion like the one that occurred in the Tunguska Valley of Siberia, Russia.

5. A light-year represents the distance traveled at the speed of light in 365 days. At five times the speed of light, traveling to the next solar system that's seven light-years away would take approximately 951 days $(5*2*44 + 7*365/5)$ or 2.6 years. Approximately, 220 days would have to be spent accelerating at 8 g's. Travel would have to be feet first when accelerating and head first when decelerating so that blood in the cranium is always maintained.

> [Author's comment: I suspect that years of space travel would cause the human body to distort. The body would become frail, the head would expand, and the eyes would bulge from the constant internal pressure of blood. Sound familiar? Acceleration in imaginary time is always present to produce a g-force that stresses the body. Consequently, distances and travel speeds dictate limitations for living beings.]

6. Traveling throughout the galaxy certainly will impose limitations, but other factors such as sector time, barriers, and wormholes could

improve desired destinations. Chapter 6 develops arguments for the existence of sector time and a location where the speed of light can be faster. Alterations in sector time and the speed of light could allow mankind to travel greater distances.

Conclusion for Chapter 1 and 2

In conclusion, there is abundant evidence to believe that another dimension is achievable. For example,

1. Space travel in another dimension by aliens has been attained. Reports on UFOs indicate the following:

   - They change colors, they travel underwater, and they're silent and vanish in and out of the atmosphere.
   - They leave messages for people like Betty and Barney Hill with comments like, "Earth is just a stopping point on a superhighway, and that our understanding of time is 'limited'." From John G. Fuller's book *The Interrupted Journey* (1966) about the abduction on September 19, 1961.
   - They use crop circles as a navigational tool; Stonehenge was an older navigational tool and so were the pyramids.
   - Natives didn't make the lines at Nazca in South America. There are too many lines. (The chapter on Supplemental Reading discusses the Nazca Lines and pictographs.)

2. Calculations of the nine planets for gravitational pull and tangential speed show that Earth is the best candidate for attaining sublight speeds prior to intergalactic space travel.

Chapter 3

EQUATIONS FOR THE GIT CURVE

In chapter 1, a curve was introduced to show the relationship between gravity, light and imaginary time. Reference to these two curves will be called the GIT curve or gravity versus imaginary time. Another new term will be parsec or short for part of a second. The relationship between a parsec in imaginary time and a second (or minute or hour) in real time is unknown at this time. It can only be arrived at empirically. Somebody would have gone there and measure it. Its relationship to real time whether it be a one-second (or two-second) shift is inconsequential. More important is the relationship between gravity and imaginary time. The GIT curve caries a higher connotation because it will imply parameters that have been measured. For example, a spaceship will pass through Earth's atmosphere at 0.5 parsecs, through the ocean at 1.0 parsecs, and through Earth at 3.0 parsecs.

Chapter 1 illustrated that imaginary time has a plus and minus value related to a future and past connotation, respectively. In his book titled *A Brief History of Time*, Stephen Hawkins states the following: "It is possible to travel to the future. That is, relativity shows that it is possible to create a time machine that will jump you forward in time." And "the first indication that the laws of physics might really allow people to travel backward in time came in 1949 when Kurt Godel discovered a new solution to Einstein's equations; that is, a new space-time allowed by the theory of general relativity."

He also states, "Again, since time and space are related, it might not surprise you that a problem closely related to the question of travel backward in time is a question of whether or not you can travel faster than light. That time travel implies faster-than-light travel is easy to see: by making the last phase of your trip in as little time as you wish, so you'd be able to travel with unlimited speed! But as we'll see, it also works the other way, if you can travel with unlimited speed, you can also travel backward in time. One cannot be possible without the other."

The problem with traveling at light speeds is that the mass of the spacecraft increases exponentially as it approaches the speed of light. In addition, the fuel consumption is unimaginable. Every college physics student back in 1964 was given the equations so that he could crank out the increased mass for himself.

Stephen Hawkins is right. Traveling backward in time is possible, but going back to visit with your great-grandmother is impossible. As stated earlier, a parsec in imaginary time could equate to only a second in real time. So we're talking about going back in time as far as 3.0 parsecs or 3.0 seconds in real time.

The other physics equations that college students learn are the ones related to the force imposed by charged particles and magnetic poles. A force field between charged particles and magnetic poles can be calculated as attractive or repulsive, depending upon polarity, with an exponentially value as the distance changes. A force field of similar expression for heavenly bodies is nonexistent. Yes, coffee cups aren't attracted to one another! And neither are heavenly bodies attracted to each other. As we'll see in chapter 6, heavenly bodies are similar to the atomic particles that follow a prescribed orbital path(s) as defined by Niels Bohr.

I'm getting ahead of myself. So before we start with the equations for the GIT curve, I would like to digress and relate a story that happened on Midway Island back in 1961. Midway Island used to be an early warning base against the possibility of missile attacks from Russia. The navy, with reconnaissance planes that patrolled the barrier between Midway and Alaska, manned the base. The planes were four-engines, prop-driven planes that had to be towed to and from the hanger. The story begins one day when a plane was taxing on the runway and drove into a pothole. Now, these potholes on Midway Island were deep, like two feet by four feet wide. If you can picture the strut of the plane sitting in a pothole and the officer of the day scrambling to make it right, then you can imagine an exasperating situation. The area was swarming with the about fifty sailors when the OD issued an order, "Get the tow truck and pull that (expletive deleted) plane out." After the first tow bar snapped like a paper twig, the OD repeated the same (explete deleted) order. As seasoned chief, who was standing there, said to the OD who was about twenty years younger, "Sir, I have a better idea." The OD snapped, "Don't bother me now, Chief." Then the second tow bar broke. The OD said, "How many do we have left?" Someone yelled out, "Two more, sir." With all the commotion, about one hundred sailors had gathered to hear the OD yell out, "Try it again"; and once more, the chief said, "Sir, I have a better idea." After the third tow bar broke, the OD didn't want to waste the last tow bar that surely would be needed for other planes. Then he ordered the pilot to fly the plane out of the pothole. The pilot revved up the engines, and the plane spun around in a circle with one wheel still in the pothole and the rest of the plane coming dangerously close to the seawall. After proving that nothing worked, the OD meekly said, "What's your idea, Chief?" The chief finally got a chance to respond, "Well, sir, with all the excitement, about two hundred sailors are here on the runway. I suggest that we send a few to fetch some two-inch hawser ropes that are in the shed over there. We tie the ropes around the strut that's in the pothole, and we use all two hundred hands to gently pull the plane out." And that's how they saved the(expletive deleted) plane!

These following equations are from an old chief like myself, who has been away from college physics for many years. But the concept is genuine and original. It's an idea that leans toward the existence of the GIT curve. It goes beyond Einstein by creating a mathematical expression for gravity.

Starting with Newton's equation, we have force equal to mass times acceleration or

$F = m*a$ where m is the mass of the object in kilograms or

pounds and

$F = m*g$ where g is acceleration exerted by Earth's pull.

Earlier we said that heavenly bodies are like coffee cups that don't attract or repel because neither has a plus or negative value associated with it. However, different heavenly bodies have different gravitation pull. Earth's gravitational pull is thirty-two feet per second per second. The moon is less, and that's why astronauts weigh less on the moon. So the formula for gravity has been sitting in front of us for years! The gravitational expression is simply: gravitation or $g = s / t^2$, where s is distance and t is real time.

The task is to equate g to a function of imaginary time. If we let real time equal to plus and minus imaginary time, we have time or $t = \pm jT$, where T is imaginary time and j is $(-1)^{1/2}$.

In the engineering world, the expression for impedance (Z) is equal to resistance plus or minus imaginary reactance or $Z = R \pm jX$. Consequently, the concept of $\pm jT$ shouldn't be a foreign mystery to us. Then if we substitute $\pm jT$ for t in the gravity equation, we get an expression for gravity of the second order; namely, Gravity or $g = s /(\pm jT)^2$.

Since distance or s has got to be a universal constant, we can express gravity on the y-axis for dependent variable and imaginary time on the x-axis for independent variable. With an equation like $y = 1/x^2$, it can be seen that y will equal infinity when x is zero, and y will equal zero when x is infinity. When the same plot is made for gravitation, g does not go to infinity when imaginary time is zero but rather thirty-two feet per second per second. When imaginary time is infinity, g takes on the characteristics of a "Gaussian bell-shaped distribution curve" approaching zero limits between three and six sigma. The GIT curve is repeated here showing the limiting characteristics when imaginary time is minimal and maximum in the positive and negative directions.

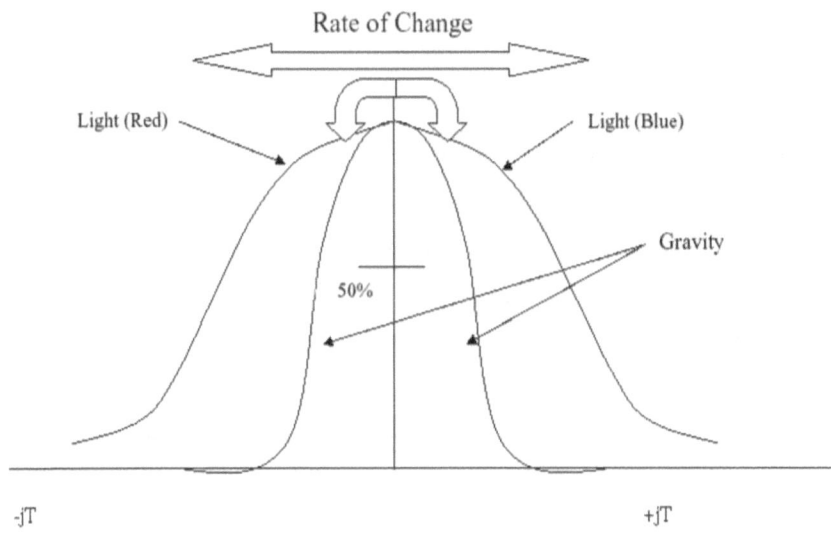

The arguments that were used to develop a gravitational curve are the same argument used to develop the light curve. Ordinary light is a function of frequency. The reds, greens, and blues all arrive at different frequencies. Basically time is equal to the inverse of 2pi times frequency where pi is approximately 3.14.

Time or $t = 1/(2*3.14*f)$

Conversely, frequency equals the inverse of 2pi times time. Substituting imaginary time and using the same arguments, we get a linear expression for frequency of the first order; namely,

Frequency or $f = 1/[2*3.14*(\pm jT)]$.

Since this is a linear relationship, the curve for frequency or light is less sever than the gravitational curve. Blue light is shown in the +jT region because it has a higher frequency than red light shown in the -jT region. I prefer to think of the frequency spectrum increasing from left to right with red on the left and blue on the right.

The arrows at the top of the curve indicate the following:

1. A changing color is associated with a spacecraft moving back and forward through imaginary time.
2. The rate-of-change signifies a spacecraft that is hovering in Earth's atmosphere.

This chapter concludes the feasibility study on proposed gravitational formulas. The next chapter investigates a method for entering imaginary time and the type of spacecraft that would be required for galactic travel.

Chapter 4

WARP SPEEDS

Warp speed is expressed as a percentage of the speed of light; for example, Warp 0.1, 1, and 10 are 10%, 100%, and 1,000% times the speed of light. The term "warp speed" was coined in the *Star Trek* TV series that started in the late '60s. The actors never explained what Warp 1, 2, 3, or etc., were because "they knew what they were talking about, and they assumed that the listener knew as well." It's human nature to assume! I can remember my calculus professor starting the session in what seemed like the middle of the book. She was talking about something that made me feel like I had missed a day in class, even though I hadn't. Again, this is human nature. This author will try to make you understand so that you don't begin to think that a page in this book is missing.

In the previous chapter, the phrase "Gaussian bell-shaped distribution curve" was used. This curve is used in statistical analysis to determine when a process doesn't meet the norm. Data is compiled into a histogram with the outer envelope forming a bell-shaped curve. Equations like sum, average, standard deviation, and etc., are calculated to indicate when a curve is abnormal. Since the section on neutrinos deals with this subject, it's important that a Gaussian curve should be understood. Here are some examples.

A scientist sets out to measure the length of all the leaves that fall to the ground underneath his maple tree in the late fall. He collects the data and learns that they all measure between 2.5 inches and 3.5 inches. He then counts the number of leaves that measured between 2.5 and 2.6 inches, then between 2.6 and 2.7 inches and etc., and all the way up to 3.4 and 3.5 inches. When he plots the number for each category against the length of the leaf, he has a histogram. The amazing thing about this histogram is that it matches the mathematical predictions of the Gaussian curve. The scientist repeats the same process with a second maple tree that is growing next to an oak tree. We'll assume that the scientist doesn't know the difference between a maple leaf and an oak leaf. As he accumulates the data and creates a histogram, he learns that some of the data is larger, and that his data is skewed. The curve is no longer centered at 3.0 inches. The scientist concludes that the process has been contaminated with another species (or another tree). If this happened at National Semiconductor, the engineer would conclude that the integrated circuit was contaminated with dust, wafers from a different yield, or a difference in the measuring technique. This is a very powerful scientific tool.

In the movie *The Time Machine* by H. G. Wells, there is a scene where a scientist designs a miniature time machine about twelve inches high. He demonstrates to his friends that the machine sitting on the table can be loaded with a cigar inside the cockpit and made to disappear by activating a small lever on the dash. His friends remarked, "Where did it go?" To which he replied, "It's sitting there in another dimension."

Using the arguments in the previous chapter, it (the prototype in the movie) wouldn't just sit there. It would be suspended in space that had been used up by Earth and its atmosphere. The arguments in chapter 2 indicated that the time machine would be suspended in a frictionless world in imaginary time as Earth sped away on its orbital trajectory around the sun. We'll review this subject again on another topic in the following chapters.

It's difficult to imagine a fourth dimension, yet alone a fifth and a sixth dimension. If these dimensions are to exist, they have to assume characteristics of their own. Radio waves or radio frequencies (RF) and photon emissions travel at the speed of light in free space. RF frequencies traveling in a conductor move at 60% the speed of light. The propagation time is reduces as RF energy moves through a different medium!

A good college book titled *Electromagnetic Waves and Radiating Systems* by Edward Jordan discusses the propagation of RF waves between two parallel plates and waveguides in chapter 7. As brief review, Jordan discusses three types of propagation; namely, transverse electromagnetic (TEM), transverse electrical (TE10), and transverse magnetic (TM10) wave forms. Basically, he is saying that TM10 mode lacks an H-field in the Z coordinate, and the TE10 lacks an E-field in the Z coordinate. He points out the TMm0 and the TEm0 modes can have multiple harmonics where m is an integer. He doesn't describe the mode of propagation after the energy field leaves the waveguide or an area between two parallel plates. Consequently, electromagnetic propagation in free space could be altered, but it still would resemble ordinary links on a chain with each link at ninety degrees to each other. The characteristic of an integer representing a harmonic of an H-field or an E-field would still be feasible. To characterize a fourth, fifth, and sixth dimension, we need electromagnetic waves with their own attributes that influence three dimensions. Refer to figure 1b the Fourth, Fifth, and Sixth Dimensions. The drawing illustrates the addition of an electromagnetic field consisting of a few links that influence a three-dimensional body. Each dimension has its own attribute because it is an extension of the basic three dimensions. The fifth dimension differs from the fourth dimension because it has more harmonic H-fields, and the sixth dimension differs because it has more harmonic E-fields.

The first section in the chapter will be dedicated toward building a time shifter. The word "time shifter" as opposed to time machine is more appropriate because the word "parsec" was defined in the previous chapter and implies a slight shift in time and not a transition in time. The second section of this chapter will be dedicated toward the construction of a galactic spacecraft capable of traveling at warp speeds.

According to historians, Thomas Edison didn't invent the lightbulb immediately. Numerous experiments with bamboo and metal filaments were tried before tungsten proved better. Similarly, a time shifter prototype has to be tried repeatedly to determine the correct field strength, frequency, and harmonic displacements.

Using Jordan's nomenclature for harmonic designation, the fourth dimension would be analogous to a TEM field where each E-fields and H-field is a fundamental frequency. The fifth dimension would be analogous to a TM10 field where the E-field is a fundamental and the H-field is a second harmonic. The sixth dimension is a TE10 field where E-field is a harmonic and the H-field is a fundamental. If E and H fields with as many as three harmonics were considered, the number of combinations would be over ten. When all the variables are considered, the number of combinations that need to be tried could exceed ten thousand. Ten different voltages for each field could be tried to generate the appropriate H-field and E-field. Ten different frequency bands could be tried along with ten different patterns. Consequently, the effort would be time-consuming with ten times ten, times ten times ten, variables or ten thousand possibilities.

Before a spacecraft could be built, the short-term goal should be the construction of a time shifter capable of producing multiple harmonics of the E and H fields that lock together with appropriate field strengths. Figure 2 illustrates a time-shift prototype model.

Construction of the Time-Shift Prototype

Construction of the prototype would start with a plastic ball maybe two or three inches in diameter. The electrostatic plates can be painted on the outside of the ball with silver conductive paint. The magnetic cores can be mounted on one-inch diameter paper rolls impregnated with iron particles. The magnet cores are glued to the outside of the ball so that stability isn't jeopardized. The circuit and battery have to be mounted on a circuit board that is positioned in the center of the ball for stability. The airstream shoots upward, floating the ball and preventing external contact. The ball has to be stable so that it doesn't wobble or gyrate up and down. The vanes attached to the outside of the ball provide a rotational of the H-field at X revolutions per second. A strobe light shines through a clear patch in the ball, striking the photocell and causing the varactor diode in the Colpitts oscillator to generate a frequency of Y revolutions per second. The experiment requires values of X and Y to be tried when they are equal and when they are multiples of each other. Experimental strength levels for electric flux measured in coulombs and magnetic flux measured in webers have to be tried. When the plastic ball disappears into imaginary time, the experiment is a success. All the harmonic values and the strength values are recorded so that a galactic spacecraft can be built with similar characteristics. We have to remember that the spacecraft will not have the initial advantage of airflow producing zero gravity.

[Author's comment: To achieve an E and H field that have equal fundamental frequencies, the values of the L and C in the oscillator have to be rather large. The varactor diode/capacitor has to be ganged with many others to produce harmonics. If fundamental frequencies between fifty and two hundred hertz can't be achieved, then a phase-shift oscillator should be tried. When simulating the fourth dimension, the fundamental frequency of the E-field has to be sinusoidal with the higher harmonics suppressed by 70 to 80 dBc. When simulating the fifth dimension, the harmonics have to be rich or plentiful as in the production of a square-wave oscillator. In both situations, the goal is to vary the fundamental between two decades; i.e., fifty to one hundred or seventy-five to one hundred fifty or one hundred to two hundred hertz.]

The May 2001 issue of *Popular Science*, has interesting discussions on a segment titled "Space at Warp Speed." On page 48, the article says, "there isn't enough mass in the universe to supply the propellant you'd need." The article describes seven different types of space drives; namely, induction sail, diode sail, differential sail, disjunction drive, induction ring, bias drive, and diametric drive. The first three space drives use photons (same as light beams) to impart energy that would propel the ship. The same concept is used in a radiometer, pictured on the cover of this book. A radiometer is an instrument used to measure the radiant energy (photons) by the torsional twist of suspended vanes that are blackened on one side and silver or aluminum painted on the other. The problem with this concept is that photons travel only at the speed of light. We want to achieve Warp 10 or ten times the speed of light. The next three space drives in the magazine discuss the feasibility of altering the properties of space. The last space drive is similar to my concept. Allow me to quote what they say:

> If a spacecraft could create an asymmetrical field around itself, somehow, then perhaps the field would propel the spacecraft. The Diametric Drive, a generic version of a 1957 "negative mass propulsion" concept, uses the interaction of a positive and negative field to propel the spacecraft.

None of these proposals have a real starting point. Consequently, we'll move on to mine that has more detail. Refer to figure 3, Design of Spacecraft's Time Shifter and Propulsion Engine. The design of the time shifter is a takeoff on an IBM patent. Back in the '70s, IBM applied for a patent that never made it into the digital world. They wanted to store ones and zeroes into some kind of a shift register. They discovered that magnetic bubbles that were part of a very thin magnetic plate could be caused to flow (or transfer) from a source, with an abundant supply, to the drive motor. This is going to sound indifferent, but back in 1950 when I was twelve years old, I remember a new game that a neighbor friend got for Christmas. It was basically a thin piece of metal that vibrated when metallic figures of football players were positioned at the line of scrimmage. The peculiarities of the vibrations caused one team to win and the other to lose.

Construction of the Spacecraft's Time Shifter
and Propulsion Engine

According to the IBM article that is no longer available to the author, the magnetic bubbles were initiated by an induction coil that produced a magnetic field across the metallic material. The induction coil is not shown in the illustration of figure 3. The oscillator driver developed a second magnetic field that was perpendicular to the first field. It caused microscopic magnetic bubbles of ones and zeroes to flow as a fast-moving stream through the plate. The concept is that a segment of the H-field can be represented as a second harmonic or "1-0-1-0-x-x-x-1-0-1-0-x-x-x" where x is either a 1 or a 0 or a void. The analogue version of this concept can be seen in figure 1b. It illustrates the H-field as a second harmonic at right angles to the E-field. This pattern is chosen over the other two choices because it's logical to think of the H-field with multiple harmonics when there are multitudes of bubbles moving in a fast stream. This fast stream allows the operating frequencies of the E and H fields to be much higher than the ones used in the construction of the prototype. For example, the E-field could operate at five hundred hertz, and the H-field would operate one thousand hertz as a second harmonic or as one thousand five hundred hertz as a third harmonic. To produce the ones and zeroes, the polarity on the induction coil has to be reversed at the same rate; namely, one thousand hertz or one thousand five hundred hertz. Push-pull circuits would provide the polarity changes to the induction coil. When all is said and done, the best feature of the spacecraft is that it has only *one moving part*!

The induction coil, the magnetic material, and the electrostatic plates are all embedded into the walls of the spacecraft with a depth of one-eighth inch. The magnetic material doesn't have to be thick. Thin electrostatic plates will provide an E-field with equal effect. The electrostatic E-field separates each fine strip of magnetic material at ninety degrees. The field strengths and harmonics of the E and H fields need to be adjusted with the knowlwdge that was achieved with the prototype. The right combinations will take the spacecraft into imaginary time. Refer to Figure 3.

Propulsion Engine

Propulsion is achieved with a neutrino trap. Speculation was made in the earlier chapters about the abundance of neutrinos and their speed of travel close to one hundred times the speed of light. Investigations into the August 2001 issue of *Discover* magazine reveal that 1,500 trillion neutrinos pass through the palm of your hand every three seconds. Scientists today are unaware of their abilities to exceed the speed of light. They're trapped in their thinking with Einstein's statement that nothing can go faster than the speed of light. Let's review what history tells us. In 1887, two scientists named Michelson and Morley performed an experiment using mirrors to reflect a light beam repeatedly, and thereby measure the speed of light. In chapter 1, the speed of Earth around the sun was calculated as 0.01% the speed of light. The scientists were concerned that the speed of Earth around the sun could influence their results. They performed the experiment at different times of the year and got the same results. Their results were contested for years until Einstein said in 1905 that their results were conclusive because the speed of light is always a constant. From that point until now, scientist were convinced that the speed of light is a constant across the entire universe.

[Author's comment: It is the author's contention that the speed of light is constant within this sector of the galaxy, but it varies to a different constant in other sectors as we shall see in the later chapters.]

Scientists don't understand how the mass of the neutrino can oscillate from being there to not being there. They give this phenomena a name called "flavor." Allow me to read the article on page 34 of the August 2001 issue of the *Discover* issue.

> The Kamioka observations also support an even stranger idea:
> that a given neutrino does not have one stable mass or one stable
> identity. Instead as it flies along, it oscillates from one identity—
> what physicists call flavor, which means a way of interacting with
> other particles—to another. A Dr. Jekyll and Mr. Hyde sort of
> affair, observed one of the Kamioka researchers.

Everyone agrees that neutrinos have zero rest mass and a slight mass when it's in motion. When it's traveling at the speed of light, it has mass measured in fractions of atomic units. However, neutrinos traveling at one hundred times the speed of light have a fractional mass that is substantially greater. The faster moving neutrino increases in mass, but at Warp 100 its mass is a fraction of the photon's mass. Within this chapter a table illustrates the propulsion force of a fast moving neutrino in comparison to the force of photon acting upon a radiometer.

Experiments deep in Earth's crust stop some of the neutrinos traveling at the speed of light, let's say Warp 1, and they stop only a few more traveling from Warp 1 to Warp 100. I read that the stoppage rate is one or two per day. In conclusion, the neutrinos have a flavor because some were traveling at Warp 1, while others were traveling at greater warp speeds with a greater mass.

Consequently, it can be speculated that neutrinos, coming from our sun, can have speeds as great as Warp 100 with a distribution resembling a Gaussian curve. With this histogram, statistics can be used to make statements like, "Ninety-five percent of all neutrinos will have a plus and minus standard deviation of 2." If our spacecraft is traveling between Warp 1 and Warp 10, then 95% of the neutrinos are available to propel the spacecraft even faster. The parallel lines in the histogram illustrate the speed of the spacecraft. The horizontal line at Warp 10 almost intersects the vertical line at two standard deviations.

Gaussian Curve for Neutrinos

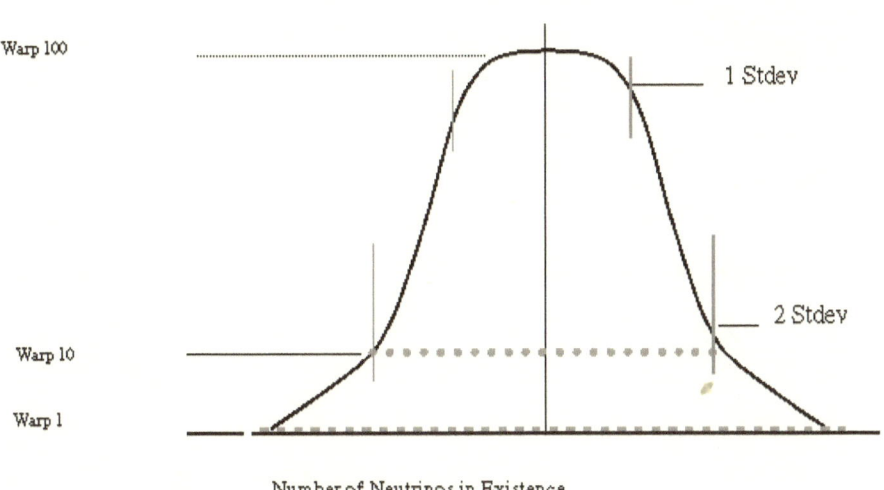

Number of Neutrinos in Existence

Stopping neutrinos on Earth is a difficult task. The same article states the following:

> "They [scientist in Kamioka, Japan] caught the particles two thousand feet down in an old zinc mine, in a cavern, lined with stainless steel and filled with fifty thousand tons of purified water."

Thomas Edison's invention for a lightbulb filament was a difficult task. The invention of a neutrino trap will be even more challenging. This doesn't mean that scientist should give up before they start when they have three advantages working for them; namely,

1. A spacecraft traveling at the initial speed of Warp 0.1, as stated in chapter 1, will catch more neutrinos than Earth traveling at Warp 0.01.

2. "The bigger they are, the harder they fall." As stated, the mass of a neutrino increases appreciably from Warp 0.01 to Warp 0.1. Therefore, the spacecraft receives improved propulsion.

3. The spaceship will accelerate from Warp 0.1 very slowly, but acceleration rate will increase more rapidly as the spacecraft transcends through Warp 1, 2, 3 up to 10 because more neutrinos will be trapped.

Neutrino Trap

The neutrino trap has to be light in weight, very thin, and dense enough to capture energy particles and thereby becoming the spacecraft's engine. Today's technology hasn't advanced far enough to suspect that a neutrino trap is feasible. It's been speculated that Christopher Columbus could *not* have built two more atomic ships to replace his wooden ones, even if he were given one sample and all the resources he needed!. However, modern man almost has the technology to build a neutrino trap that is relatively thin and capable of stopping neutrinos. Before I suggest one possibility, I have to relate this old joke about Columbus:

> Christopher Columbus started out in four wooden ships to prove that the world is round and not flat. There weren't three ships. There were four; i.e., the Nina, the Pinta, the Santa Maria, and . . . and . . . and? Nobody remembers that name of the fourth ship because it sailed off the edge!

The *Chicago Tribune* reports in the January 31, 2008, newspaper that scientist at Northwestern University in Illinois used DNA to build 3-D nanostructures.

The article states, "Nanoscientists hope they can use basic elements such as gold, silver, carbon and others to make materials that are harder, more heat resistant or better electrical conductors than anything produced by nature. The chief chemist said, 'The dream is to learn how to break everything into elemental building blocks and then assemble particles into preconceived architectures. If you have that kind of control, in principal you can build a material with any property you want.'"

Train and Bullet Theorem

The best way to explain neutrino propulsion is with easy math and the assumption that every variable has a linear relation. The goal is to explain how neutrinos can develop enough force to propel a spacecraft when their mass is less than a photon. The best way to explain this phenomenon is with an analogue called the "train and bullet theorem." For example, a man with a rifle is standing at the back of a train and asked to hit a stationary hardwood target on the caboose. He shoots, and the bullet penetrates the hard wood ten centimeters within a fraction of time called t^1. The event time or t^1 is the difference between the time when the bullet entered the hardwood and when it stopped. In the second situation, the train is traveling past the rifleman at one-tenth the speed of the bullet. The rifleman shoots at the target again, and the bullet penetrated the hardwood by only one centimeter. If the bullet was traveling at the same speed, then the event time is one-tenth of the original measurement. The two situations can be mathematically expressed as

$$V \text{ (velocity)} = 10 \text{ cm (distance)} / t^1 \text{ (time)} = 1 \text{ cm (distance)} / t^1 \text{ (time)} / 10 = 10$$

From this equation, it is easy to see how the event time or t^1 decreases proportionally as the train moves faster. When calculating the neutrino force, the bullet becomes the neutrino, and the train becomes the spacecraft. The target becomes the neutrino trap. To understand how propulsion is feasible at Warp 0.1, we have to compare the force that a radiometer experiences with the force that a neutron trap experiences at Warp 0.01, 0.1, 1, and 10. In the earlier chapters, we calculated Warp 0.01 as Earth's speed and Warp 0.1 as the spacecraft's breakaway speed. Let's assume the following:

- The radiometer on the cover of this book is driven by a work force of 10 units, as per the equation $F = m*a$ or $Ft^1 = m*v$ where t^1 is the event time.

- The mass of the photon striking the vane is normalized to $10*m^1$. We could use the real weight that is a fraction of an atomic unit, but we don't need to do so.

- The velocity is the speed or light normalized to one (1) or $1*c$.

- The event time or t^1 is normalized to one (1) as Earth travels at 0.01% the speed of light. The event time t^1 is the duration that the force lasts. In the case of the radiometer, a photon with a mass of 10 (units) hits the vane at light speed, penetrating the surface of the vane and sinking to a level within 1 x-seconds with a force of 10 units. The x can be micro, nano, or pico. The mathematically expression is $F = 10m^1*c/t^1 = 10$ units.

- As per the "train and bullet theorem," the event time decreases by units of 10 as the speed of the spacecraft increases by units of 10.

- The mass of the neutrinos striking Earth at Warp 1 is equal to $m^1/1000$.

- The mass of the neutrinos striking Earth at Warp 10 is equal to $m^1/100$.

- The mass of the neutrinos striking Earth at Warp 100 is equal to $m^1/10$.

- The direction of the spacecraft is sailing "downwind" of the neutrinos.
- The neutrino trap is a thin sheet of material made from nanostructures.

In reality, the relation between mass and speed is not a linear equation as we have assumed but rather an exponential relationship. For demonstration purposes, we need to illustrate how a neutrino has no impact on Earth and yet can develop adequate force to propel a spacecraft.

Propulsion Force Table Using the Equation F = m*c/t

Time (t^1)	$1*t^1$	$.1*t^1$	$.01*t^1$	$.001*t^1$
Warp Speed of	Warp .01 Earth	Warp .1 Spacecraft	Warp 1 Spacecraft	Warp 10 Spacecraft
Photon Force @ Radiometer @ Warp 1	$F=10m^1*c/t^1$ = **10 units**		Spacecraft faster than primary photon (See Chapter 1)	Spacecraft faster than primary photon (See Chapter 1)
Neutrino Force @ Trap @ Warp 1	$F=(m^1/1000)*c/t^1$ =.001 unit undetected	$F=(m^1/1000)*c/.1t^1$ =.01 unit	Spacecraft faster than neutrino	Spacecraft faster than neutrino
Neutrino Force @ Trap @ Warp 10	$F=(m^1/100)*10c/t^1$ = .1 unit undetected	$F=(m^1/100)*10c/.1 t^1$ = 1 unit	$F=(m^1/100)*10c/.01t^1$ = **10 units**	Spacecraft faster than neutrino
Neutrino Force @ Trap @ Warp 100	$F=(m^1/10)*100c/t^1$ =10 units barely detected	$F=(m^1/10)*100c/.1t^1$ = **100 units**	$F=(m^1/10)*100c/.01t^1$ = **1000 units**	$F=(m^1/10)*100c/.001t^1$ = **10,000 units**

The table illustrates the relative force of a neutrino with a smaller mass than a photon and the anticipated force required driving a radiometer at 10 units. The table indicates that many photons can be detected with a radiometer on Earth traveling at Warp 0.01, and only a few neutrinos could be detected with a special nanostructure as a neutrino trap. When the spacecraft accelerates to Warp 0.1, an adequate number of neutrinos are available to start propelling the spacecraft. When the spacecraft travels at Warp 1 and Warp 10, the force increases to 1,000 and 10,000 units, an amount considerably greater than the force required to drive the radiometer. The table demonstrates feasibility with the neutrino at 1/1,000 and 1/100 the mass of the photon.

Conclusion for the Propulsion Force Table

The propulsion force table is meant to illustrate that neutrinos smaller in mass can have more force than a photon when the spacecraft is moving faster than Earth. In retrospect, it is suspected that more accurate calculations will show that the neutrinos traveling at Warp 1 to Warp 10 are more effective accelerators than the neutrinos traveling at Warp 100.

Conception of a Neutrino Trap

A neutrino trap intended for space travel should consist of a laminated device four to eight inches thick and constructed of either the cellular type or the electron type material. The cellular type is a compound that exists in a semiliquid and a semisolid state. Onion skin with its many cells is a compound that exists in a semiliquid and a semisolid state. Consequently, a cellular neutrino trap would have solid characteristics and contain a liquidlike core. Atoms within the core exist as ions. Ideally, a cellular trap would look metallic like tin foil, have one hundred cells or more per inch and the ability to bounce after it had been rolled into a tight ball. This section will develop the feasibility of two types of neutrino traps; namely,

- the cellular neutrino trap and
- the electron neutrino trap.

In the cellular type, the neutrino transfers its speed to the ion. A multiple number of ions create the force that pushes the spacecraft. In the electron type, an electron transfers its charge to the neutrino. A multiple number of charged neutrinos create the force that pushes the spacecraft. The concept for both types is difficult to describe on the subatomic level. However, the idea can be conveyed in an analogous manner. Imagine a neutrino like a cue ball entering a pool table. The pool walls offer no restriction to the neutrino or cue ball, but an object called an ion does. The ion can be thought of as a racket ball with an elasticity that exceeds the cue ball. In the analogue world, the cue ball enters the pool table and strikes the racket ball. The racket ball moves much faster than the cue ball because of its elasticity. It bounces off the rubber bumpers releasing a vector force and then collides with the cue ball a second time. The speed of the cue ball slows down and continues to the next pool table. The racket ball bounces off the rubber bumper on the other side releasing a vector force again. (The term "vector force" is used to describe the components of force in the x direction and the path that the spacecraft is traveling or the y direction.) In the subatomic world, the ideal ion is lighter and more elastic that the neutrino. The neutrino collides with the ion in a cell that is less than 0.01 inches square. The ion bounces between cell walls releasing vector forces. The process is repeated each time after the neutrino enters the next cell. The vector force in the y direction is what drives the neutrino trap and spacecraft.

A secondary goal for the cellular neutrino trap is to generate electricity. If a conductive compound is used on a small portion of the trap, an electric current can be created similar to a photovoltaic cell. Neutrinos replacing photons could be used as an alternative form of energy within the spacecraft and on Earth as power plants. They could replace coal burning power plants and contribute zero pollution to the environment. When used in cars, there wouldn't be any need for gasoline.

The electron neutrino trap is constructed of a device that has an abundance of electrons flowing in-line but at a slight angle to the direction of the neutrinos. A device of this type would have to contain a cathode and an anode within a sealed structure. The mechanics of the propulsion start with a neutrino passing through an electron with a slight deflection toward a larger positive-charged plate. The size of an electron with respect to a neutrino is the same as the size of a grapefruit is to apple seed. The neutrino assumes a negative charge because it passes through a huge negative field created by the electron. A multiple number of negative-charged neutrinos strike a positive plate releasing its vector force. Some neutrino try to pass through the plate, but the force of attraction alters their course back to the plate. Each time the charged neutrinos strike the plate, they release their vector force. The vector force in the y direction is what drives the neutrino trap and spacecraft.

[Author's comment: The author regrets that he can't use drawings to convey these two different concepts. It is his belief that a nanostructure might meet the requirements of the cellular neutrino trap.]

Detail Design of the Neutrino Trap

If I were a microbiologist and asked to design a neutrino trap, I would have difficulty finding a starting point. Neutrinos originate on the sun and penetrate the Earth from one side to the opposite side. They are detected hundreds of feet below the ground.

Before starting this quest, we have to remember that there is a difference between "detecting a neutrino" and "using a neutrino". For example, I can "detect" where my car is in a parking lot with the remote horn control located on my key chain, but I will "use" my car by climbing inside and driving away. Consequently, the goal is to "use" the energy contained within a stream of neutrinos—a task that's much easier.

I would start with a request to construct a neutrino generator on a microscopic scale. The engineers at Fermi Labs in Batavia, Illinois already have one. I would need a generator that would be mounted under a microscope on a rotating gimbal that is always perpendicular to the rays of the sun. This instrument would provide a visual measurement when a slide with an experimental substance is activated by neutrinos. The task is to place a substance on the slide, radiate it with neutrinos, and observe any physical change in position. Of coarse, this isn't going to be an easy task. It requires repetition with a multitude of substances. To create a controlled experiment, the natural emanation of neutrinos from the sun has to be ruled out. This is achieved by using a controlled supply of neutrino pulses that strike the test substance in a direction that is perpendicular to the sun's rays. This sophisticated instrument will be referred to, as simply "the test bed".

If I had to look for an appropriate substance, I would start with spider silk. A news item in the July 27, 2008, issue of the Chicago Tribune states the following:

> "Ms. Cheryl Hayashi keeps black widows, tarantulas, and jumping spiders, to name a few, at her laboratory at the University of California, Riverside. Last fall, the associate biology professor's promising work on spider silk helped her win a $500,000 'genius' grant from Chicago's John D. and Catherine T. MacArthur Foundation."

In another article dated November 2005, Discovery magazine documented the work that Ms. Cheryl Hayashi and Ms. Jessica Garb did at the University of California.

> "Some things about spider silk are difficult to understand. But Garb and her colleague Hayashi have unraveled the genetic code for one of the more perplexing types of silk—the strands that spiders use to weave their egg cases. Each case must be tough enough to keep out parasites, impermeable to rain and fungus, and breathable while insulating eggs from temperature extremes. Garb and Hayashi speculate the egg cases may even **block ultraviolet light**."

If spider silk around the egg cases is lightweight, stronger than steel, and blocks ultraviolet light, I would start my experiments with the silk of a spider web or egg case. The theory is that spider silk and egg cases blocking ultraviolet light, also block neutrinos. It's logical to think that the spider evolved with the production of an egg case that protects its young when exposed to sunlight. It's also logical to assume that evolution produced an egg case that blocks neutrinos. As discussed earlier, photons (light rays) and neutrinos both have "zero" rest mass. When they are traveling through space, they have an infinitesimally small mass—smaller than the mass of an electron. A stream of photons called packets travel in a spiral frequency, something like

"eeee eeee eeee eeee",

and neutrinos travel in a straight line, something like

"------------------".

Consequently, the propagation of photons throughout space is easy to stop, but the stoppage of neutrinos requires a more elaborate plan.

In a photovoltaic cell or PV, photons knock electrons loose. This is equivalent to a Volkswagen being moved by a stream of Ping-Pong balls. On a molecular scale the following happens:

"The process of adding impurities on purpose to pure silicon is called **doping**, and when doped with phosphorous, the resulting silicon is called **N-type** ("n" for negative) because of the prevalence of free electrons. N-type doped silicon is a much better conductor than pure silicon is.

Actually, only part of the solar cell is N-type. The other part is doped with boron, which has only three electrons in its outer shell instead of four, to become **P-type** silicon. Instead of having free electrons, P-type silicon ("p" for positive) has free holes. Holes really are just the absence of electrons, so they carry the opposite (positive) charge. They move around just like electrons.

The interesting part starts when you put N-type silicon together with P-type silicon. Every photovoltaic cell has at least one **electric field**. Without an electric field, the cell wouldn't work. This field forms when the N-type and P-type silicon are in contact. Suddenly, the free electrons in the N side, which have been looking all over for holes to fall into, see all the free holes on the P side, and there's a mad rush to fill them in."

The result is the conversion of light into electricity. To obtain motion the goal would be to convert the energy of neutrinos into the same useful product. Again, it's the same analogy as using Ping-Pong balls to move a Volkswagen. So let's make this task easy and assume the spider silk contains a substance that blocks neutrinos. The experimental task is to take a sample, position it on a slide under the microscope on the test bed, radiate it, and observe if there is any movement. If success isn't achieved, we move on to sampling the cone in a beehive or the substance of a wasp's nest.

Let's assume that success is achieved. What do we do with the substance? To build an engine, the substance needs to be combined with printer's ink. Imagine a page like this one with each letter capable of stopping a neutrino or slowing it down. Some letters like "I, L, and H" would be more productive in trapping neutrinos coming from an angle that's left or right of center. Other letters like "B, E, and F" would do just the opposite. If the activation of specific letters is controlled electronically like the pixels on a Liquid Crystal TV monitor, then a rudimentary form of steering can be achieved. Now imagine a multiple number of pages (or book) like the first one with letters that provide a thrust into three-dimensional space. You could hold this processed book with the cover facing the sun and the book would move you in a zero degree radial direction very fast, a thirty-degree radial direction somewhat slower, and a ninety-degree radial direction with zero speed. If each letter has the capability of "raised" movement above the paper, causing the paper to follow, then the chance of capturing every neutrino is very good. The processed book is to the sun as a sail is to the wind. This book contains approximately 200,000 characters or letters. The chance of a particle penetrating this book without striking one letter is minimal. Every letter would represent a cell capable of movement with the constant bombardment of neutrino streams.

What happens if the spider substance and ink combination doesn't work? Is there a plan B?

Plan B would involve the construction of microscopic cells with the ideal substance inside the cell. In earlier paragraphs, reference was made to a cue ball striking a racket ball analogous to a neutrino striking an ion. The cell was described as an elongated rectangle. In reality, the cell would have to be shaped like a golf ball sitting on a tee. When the neutrino enters the cell that is shaped like the golf ball, it strikes the ion pushing it toward the back of the cell shaped like the tee. With the ion traveling down the tee part, it gets bombarded over and over again by a stream of neutrinos. Each time it is hit, it transfers the energy of the neutrino into the walls of the cell causing movement until it bumps into the wall at the tip of the tee. The goal is to make the ion elastic and as light weight as possible. If the ion is elastic, energy in the form of acceleration will be given to the ion making it travel faster than the speed of the neutrino. More collisions will occur if the ion is elastic and if the ion is forced to travel down the tee section of the cell that is shaped like a funnel. The ion will have multiple collisions if it always travels farther for the same length of time. The distance traveled by neutrino and ion are as follows:

$s(n) = v*t$ where $s(n)$ is the distance that the neutrino travels and
$s(i) = v*t + a*t^2$ where $s(i)$ is the distance that the ion travels
with v = velocity and a = acceleration and t = any given time period.

To recycle the process repeatedly, an E-field has to be established with the golf ball part of the cell charged positive and the tee part of the cell charged negative. The ion has an initial charge to begin with, but it picks up an additional charge each time it strikes the cell wall. At the end of the cycle, the positive voltage is applied and the ion drifts back to its original starting point. The energy required to relocate the ion is always smaller that the energy that the neutrino released during the collision process.

The cells of neutrino trap described under plan B are similar to the pixels provided by a liquid crystal display or LCD in a TV monitor. Each pixel is controlled by a voltage that allows colored light at a specific wavelength to pass through with the source being a white light in the back of the TV. A matrix screen (or grid) that is composed of transparent wires supplies the voltage applied to each pixel. A Bingo card functions the same way. The Bingo number is selected with a designated row and column. Consequently a neutrino trap of this nature would have as many tiny cells as the pixels on a TV monitor and operate under a matrix-voltage control.

In the search for an ideal substance, many experiments would have to be made on the test bed to find a nanostructure that has these characteristics. Perhaps a TV pixel mounted on the test bed would reveal some interesting results in blocking neutrinos!

If the plan for developing a neutrino trap described in original plan (or plan A) is a success, it could easily be converted into a power generator for supplying electricity. One of the basic rules that an electrical engineer learns is that there are three variables closely related; namely, motion, current, and field (either E-field or H-field). If you put current into a motor that has a permanent H-field (or create an electromagnetic field), motion is produced. If you put motion and an H-field into a system, you create a voltage generator. To create a power plant using neutrinos, the intermediate steps would be motion and static electricity. We have described a processed book made with special ink that causes the letters to move against the pages. The same processed book could be used as a power plant to create electricity. Allow me to quote an article.

"The **relative position** of two substances in the triboelectric series below tells how substances will act when brought into contact. Glass rubbed by silk causes a charge separation because they are several positions apart in the table. The same applies for amber and wool. The farther the separation in the table, the greater the effect.

When two non-conducting materials come into contact with each other, a chemical bond, known as **adhesion**, is formed between the two materials. Depending on the triboelectric properties of the materials, one material may "capture" some of the electrons from the other material. If the two materials are separated from each other, a **charge imbalance** will occur. The material that captured the electron is negatively charged and the material that lost an electron is positively charged. This charge imbalance is where "static electricity" comes from. The term "static" in this case is deceptive, because it implies "no motion," when in reality it is very common and necessary for charge imbalances to flow. The spark you feel when you touch a doorknob is an example of such flow."

- Human hands (usually too moist, though) *Very positive*
- Rabbit Fur
- Glass
- Human hair
- Nylon
- Wool
- Fur
- Lead
- Silk
- Aluminum
- Paper

- Cotton
- Steel *Neutral*
- Wood
- Amber
- Hard rubber
- Nickel, Copper
- Brass, Silver
- Gold, Platinum
- Polyester
- Styrene (Styrofoam)
- Saran ® Wrap
- Polyurethane
- Polyethylene (like Scotch Tape)
- Polypropylene
- Vinyl (PVC)
- Silicon
- Teflon *Very negative*

Ideally, a power plant would consist of special ink-letters printed on Teflon sheets in such a way that the Teflon rubs against nylon sheets when the letters move. The Teflon and nylon sheets are alternated as the even an odd pages in a book. In this case the processed book doesn't produce motion but rather electricity. To comply with basic engineering rules, additional motion would have to be added to the book. Motion goes in; neutrinos go in; and out comes electricity.

[Author's comment: Preliminary experiments at home indicated that I could create one quarter volt of pulsed AC voltage with only six sheets of Saran ® Wrap folded between the sheets of a thick paper book. Each time I moved the top cover of the book with respect to the stationary bottom cover, I witnessed positive and negative voltage excursions of plus and minus 250 millivolts on an oscilloscope.]

In conclusion a power plant could be designed with a multitude of moving cells as opposed to six sheets of Saran ® Wrap. Each cell would be a special ink-letter adding its static voltage to the next. The AC voltage would have to be rectified and a small amount of current would be required by a DC motor to provide motion for the book. The rest of the rectified current could be used to power the electrical requirements of a spacecraft or even a hybrid automobile.

[Author's comment: The thought of an alternative fuel source for automobiles that is environmentally friendly is beyond belief. The thought of automobiles passing though an intersection without any traffic lights is science fiction. Why not a reality? This book is dedicated to spacecraft traveling through imaginary time and not crashing into physical objects. So why can't automobiles do the same? The benefit is the reduction in the price of oil and the modification in mass transportation. Without any traffic lights or any other autos to crash into, a typical forty-five minute trip to work at an average speed of thirty miles per hour would take only fifteen minutes, traveling at ninety miles per hour.]

The benefits in discovering a new frontier called "imaginary time" are overwhelming. The benefits in utilizing the power of the neutrinos for space travel and commercial applications are unimaginable, but practical. In the future mankind will need to procreate its species across the galaxy. With population growth and overcrowding, mankind will need improved transportation on land and throughout space. The benefits of all these advantages have a logical starting point. The starting point begins with this book and excites the imagination of others. You—the reader—will provide these benefits! Or you—the reader—will derive these benefits sometime in the future!

Location of the Neutrino Trap and Navigation

We'll call the neutrino trap an engine even though it is a sail mounted on the ceiling of the spacecraft. Every experienced sailor knows that a sail has the same curvature as the top portion of an airplane wing. The speed of the air that leaves the sail of foil is greater than the speed of the air that enters. The difference provides thrust and more important maneuverability. This author suspects that Columbus had navigational guidance system that was never mentioned in his journals and was somehow lost. Prior to the onboard latitude clock, Columbus was able to sail in a straight line to a longitude where India should have been—a place where he could improve trade. With time running out, he had to take the straightest path to cover the three-month journey. The author suspects that he used two ships to tack and turn in a rhythmic motion to create a straight line. With Columbus in the third ship, he had a relative view of the other two and could plot a straight course. If this concept existed in 1492, perhaps it can be used for future guidance systems. With every star in the galaxy shedding streams of neutrinos, the possibilities of crosscurrents arise. Consequently, the pilot of the spacecraft has to tack and turn as he sails through the galaxy.

When sailing through the Earth's atmosphere, the pilot has to tack and turn at a forty-five degree angle with respect to the directional flow of neutrinos from the sun. Each forty-five degree adds to produce a visual effect from the ground that the pilot is making unearthly movements of ninety degrees.

Navigational guidance and landing present additional problems that need to be resolved. An ordinary gyroscope would probably expire on a journey to another solar system seven light-years away. Navigation with a tack-and-turn approach would be required to sail toward a star that was emitting neutrinos head-on.

Traveling at Warp 10 and trying to land on a planet is like running toward a merry-go-round and trying to jump on. To achieve this goal, the space traveler has to leave the imaginary time world that is frictionless and enter the real time world, using the gravitational pull of the planet and the friction of its atmosphere (if it has one) to slow it down. After this is accomplished, he has two choices. He can land either at night or in the daytime. There are problems with either choice. If he lands at night, he is approaching Earth from an area in space that was recently occupied. At night, Earth has a "wake side" like the water trail left by a speedboat. He can helicopter down by shifting back and forth between real time and imaginary time, but visibility is limited without runway landing lights. The other choice is to land in the daytime. The tangential speed of Earth around its axis is one thousand miles per hour. The tangential speed of Earth around the sun is sixty-six thousand miles per hour, much faster than merry-go-round speeds. Helicoptering down during the daytime is impossible because the spaceship would be between Earth and the sun, and switching to imaginary time would cause the spacecraft to lose the same orbital speed around the sun that Earth has. The only way to accomplish a visual landing in the daytime is to do a "forced" landing. A forced landing occurs when the pilot enters real time, approaches his chosen landing site, and discoveries that it's spinning in the wrong direction at one thousand miles per hour. How does this happen? Well, the sun rises in the east and sets in the west, and the pilot always has to approach the planet from the wake side, making his direction of travel opposite to the spin of the Earth—a small penalty to pay when jumping in and out of imaginary time.

In the past, visitors from outer space paid this penalty every time they landed. Evidence exists in Earth's scars that were caused by alien visitors. Additional information on this subject is included in chapter 6 with reference to photographs in chapter 8.

Worse-Case Scenario

Monumental problems can occur with landings made by a novice pilot. Let's assume a worse case scenario happens something like this in the distance future:

1. The year is AD 2200, and you are coming home from a business trip on another solar system.
2. You don't live on Earth, but the pilot has to fly past our solar system to get to yours.
3. The pilot plans to execute a slowdown so that the passengers can see the Blue Planet.
4. The pilot is a novice and always confuses the time-shift handle with the neutrino speed-control handle.
5. You are a passenger traveling at Warp 5 in imaginary time and unaware of what can befall you.
6. The pilot has a chart like the one in figure 1a and calculates that he has 8.7 minutes to reduce his speed from Warp 5 to sublight after he passes Jupiter's orbit. If he were flying at Warp 20, he would have 2.2 minutes. [We're assuming that the problems of deceleration time have been resolved.]
7. The pilot sees Jupiter flying past and immediately engages the time-shift handle, launching the spacecraft from the protective environment of imaginary time to the harsh environment of real time.
8. Small asteroids in the asteroid belt and stardust are bouncing off the hull of your spacecraft slowing it down.

9. Space radiation that never was a problem in imaginary time is now contaminating the passengers with gamma poisoning.

10. The amount of gamma radiation that the pilot absorbs flying at Warp 5 is overwhelming and incapacitating him quickly.

11. In his delirium, the pilot tries to reverse a bad decision by opening the grate on the neutrino trap. This is the decision that he should have made first. But it's too late.

12. The spacecraft is rushing toward Earth with less than 8.7 minutes to impact!

13. The spacecraft's speed is reducing from Warp 5 to Warp 2 as it enters Earth's atmosphere. According to Einstein's equations, the mass of a spacecraft traveling at Warp 5 in real time is equal to infinity plus some other number. Let's say X. As the spacecraft slows down, its mass is equal to infinity minus some number. Let's say Y.

14. The Earth experiences the change in mass like the critical mass of plutonium decomposing.

15. The released energy is per the formula or $E = [X-(-Y)] * c^2$, causing a nuclear explosion twenty thousand times greater than the atomic bomb dropped over Hiroshima.

16. Your business trip is over! Ended! Your boss will be mad as hell that you didn't stick around to complete the assignment. Bosses are like that! You're dead, and your boss wants to fire you. Ahh, he'll probably fire you anyway!

17. The people on Earth will say, "Oh, look. It's another meteor like the one that occurred over the Tunguska Valley of Siberia, Russia." Or, "Yea. It's similar to the fire ball that landed in Southwest Egypt, heated the sand, and turned into opaque glass crystals without leaving any sign of a crater."

Physics Equations Explained

Most physicists will have trouble accepting the feasibility of imaginary time. They are conditioned to understanding equations like the Lorentz transformations and assuming that a number like infinity is meaningless. Einstein deduced that space and time coordinates used by each individual observer are interconnected by formulas of transformations known under the name of Lorentz transformations. The Lorentz equation indicates the traveler's time with respect to the observer's time will slow down becoming infinitely large as the traveler's speed equals light speed. At the speed of light, the equation equals infinity or $1 \div 0$ (or 1/0). This equation is the reason why scientists believe that the speed of light can't be exceeded. The Lorentz equation states,

$$t_o = (t - \beta x/c) / (1 - \beta^2)^{\wedge \frac{1}{2}} \qquad \text{where } \beta = v/c \text{ and}$$

v is spacecraft's velocity and

c is the speed of light.

t_o is proper time or Δt_o or expanded time for the traveler.

t is the time experienced by the observer on Earth with a change of Δt equal to $(t - \beta x/c)$.

x is the distance traveled along the x-axis.

If we work with this equation at a Warp speed of 5, we have:

$$t_o = \Delta t_o = (\Delta t) / (1 - 5^2)^{\wedge \frac{1}{2}}$$
$$= (\Delta t) / [(-1)^{\wedge \frac{1}{2}} * (-1 + 5^2)^{\wedge \frac{1}{2}}]$$
$$= (\Delta t) / [\, j*(-1 + 5^2)^{\wedge \frac{1}{2}}] \text{ where } j \text{ is imaginary time or } (-1)^{\wedge \frac{1}{2}}$$
$$= -j\, \Delta t / (25 - 1)^{\wedge \frac{1}{2}} \text{ where } 1/j = -j \text{ and } j * -j = 1$$
$$\cong -j\, \Delta t / 5 \text{ a time difference of 2 years if } \Delta t \text{ is 10 years}$$

In this case the Earthling says to the traveler upon his return, "You've been gone ten years." The traveler responds with, "No. I've been gone only two years."

In the worse-case scenario, your spacecraft crashed. Now we'll create a scenario where you're traveling past Earth on your way home to see your twin brother.

1. You sail past the Blue Planet in *negative imaginary time at Warp 5*.
2. The planet Earth doesn't look blue; it looks green. See chapter 1.
3. You're anxious to see your twin brother.
4. You've noted that you've been gone for ten years. You and your brother will be ten years older as per your internal biological clocks. Depending upon your speed of travel, your twin brother will look older when you get back home.
5. If your spacecraft had been traveling at Warp 0.5 or one-half the speed of light, the difference in aging between you and your brother would have been 1.5 years. Derived from the equation $t_0 = 10 / [1 - (1/2)^2]^{\wedge 1/2} = 11.5$ years.
6. If you had been sailing at Warp 5, you would have been sailing in negative imaginary time as per the equation with the symbol "**-j**". The time difference between your clock in the spacecraft and your brother's clock in his house would be two years, as per your perspective. You won't be two years younger when you get home. You went back in imaginary time, not real time! Consequently, your brother would look two years older.
7. If your spacecraft had been traveling at Warp 0.75 or three-fourths the speed of light, then the difference in looks between you and your brother would have been 5.1 years. Derived from the equation $t_0 = 10 / [1 - (3/4)^2]^{\wedge 1/2} = 15.1$ years.

The following table summarizes how much older your twin brother would look compared to your image. Smaller age differences may not be noticeable other than a few extra white hairs. When your spacecraft traveled faster at Warp 5, you aged 10 years, and your brother aged 12 years; that is, biological years. When your spacecraft traveled at Warp 0.75, you aged 10 years and your brother aged 15.1 years. Doesn't this mean that you traveled back in time by going faster as we proposed in the earlier chapters? If this is the case, then the pilot's instructions should state that the spacecraft should always accelerate between Warp 0.75 and Warp 1 as quickly as possible and travel at Warp 5 or faster.

Time Factors at Warp Speeds

Warp Speed (%)	Warp Speed (by Name)	Expanded Time (years)	Your brother looks
50	Warp 0.5	11.5	1.5 years older
75	Warp 0.75	15.1	5.1 years older
80	Warp 0.80	16.7	6.7 years older
90	Warp 0.90	22.9	12.9 years older
95	Warp 0.95	32.0	22.0 years older
500	Warp 5	2.0	2.0 years older
1,000	Warp 10	1.0	1.0 years older
2,000	Warp 20	0.5	0.5 years older
10,000	Warp 100	0.1	0.1 years older

Acceleration and G-force

This section is dedicated to the properties of outer space that are unrelated to known physics laws that exist today. Before we can speculate about the conditions of outer space, we need to establish a corollary with known properties here on Earth. Observation is our most important tool for predicting the parameters of space travel into the future. Let's study some examples.

There is a legend that Christopher Columbus had beliefs that were contrary to public opinion. Almost everyone believed that the world was flat. Kings and queens were told by leading scientist that the world was flat. Let us ask how Columbus formed a different opinion. According to the legend, he would stand on the coastline and observe how ships decrease in size and amplitude as they slowly sailed into the horizon. Others agreed to those observations and concluded that eventually they would sail over the edge by continuing in the same direction. Their proof was that many ships never came back. Columbus concluded that the ships weren't sailing off the edge but rather, crossing a world that is round. The body of the ship would disappear first and then the top of the sail was always the last to go.

With similar logic, we're going to show first, that a g-force of 8 g's has a way of reducing itself, and secondly, that communication to the home planet is feasible when traveling at warp speeds. In the earlier chapters it was shown that an acceleration of eight times one g-force was necessary to achieve Warp 1 within 44-days. To accelerate from Warp 1 to Warp 2 in another 44-days, a constant acceleration of 8 g's was still required. The problem is that the human body can not sustain large g-forces over a long period of time without supplemental equipment. Consequently, the first chapters were written with the concept that the human body could endure with special equipment yet to be discovered. We had to assume that visitation to other solar systems five to fifteen light years away was possible within a reasonable length of time. Figure 1a indicated that a solar system located 7 light-years away could be reached in 2.6 years by traveling at Warp 5. We also assumed that the upper limit was Warp 10. But is it? If neutrinos are traveling at Warp 100, is there a limit as to how fast a spacecraft can travel?

It does get better. It has to. It would be too ironic if mankind couldn't travel to distant stars because of physical limitations. So let's postulate a new approach. On page 71, discussion was made about the shrinkage of time using the Lorentz transformation formula. If traveling throughout the galaxy is feasible, then we're going to need another formula that shrinks the g-force. Fortunately, we should be able to extend the Lorentz formula to include g-force and then shrink it. This should be possible because acceleration and g-force are still a function of time.

First, we have to distinguish between acceleration and g-force. Like Columbus, we'll make some observations. Let's assume that a specially designed racecar can accelerate to five hundred miles per hour. The back cushion is built with a special foam rubber, twenty inches thick to absorb the acceleration shock. The driver climbs into the racecar and accelerates from 10 miles per hour to 100 miles per hour and notice that he sinks into the back cushion by 10 inches. With the same acceleration he accelerates from 100 miles per hour to 200 miles per hour and notices that he sinks into the back cushion by 2 inches. With the same acceleration from 200 to 300 miles per hour, he sinks 1.5 inches. What he's experiencing is a g-force shrinkage factor! So let's define $a(v)$ as the forward acceleration of the vehicle, $g(f)$ as the forward g-force, and $g(e)$ as the gravitational pull of the Earth. In the earlier chapters, the $g(e)$ factor was eliminated when the spacecraft made the transition into imaginary time. The downward pull of the Earth, $g(e)$, at 32 feet per second per second no longer enters the equation. As indicated on page 71 the Lorentz equation states:

$$t_0 = (t - \beta x/c) / (1 - \beta^2)^{\wedge \frac{1}{2}} \qquad \text{where } \beta = v/c$$

Since acceleration and gravitational pull are related to time, the same formula can be extended to include them as follows.

$$g(f) = a(v) / (1 - \beta^2)^{\wedge \frac{1}{2}} \qquad \text{where } \beta = v/c \text{ and } \beta = \text{warp speed or } w$$

The symbol w is the ratio of the speed of spacecraft divided by the speed of light with velocity units that cancel each other. After substituting w for β, we can rewrite the equation as follows:

$$g(f) = a(v) / (1 - β^2)^{\wedge\frac{1}{2}}$$
$$g(f) = a(v) / (1 - w^2)^{\wedge\frac{1}{2}}$$

The symbol for imaginary time is j where j = (-1)$^{\wedge\frac{1}{2}}$. The transformation then produces:

$$g(f) = a(v) / (-1)^{\wedge\frac{1}{2}} * (-1 + w^2)^{\wedge\frac{1}{2}}$$
$$g(f) = a(v) / j * (w^2 - 1)^{\wedge\frac{1}{2}}$$

With the spacecraft traveling in imaginary time, the j factor can be eliminated.

$$g(f) = a(v) / (w^2 - 1)^{\wedge\frac{1}{2}}$$

With large warp speed values, the formula can be reduced to the following.

$$g(f) = a(v) / w \qquad \text{where w is warp speed}$$

This is a significant equation that applies to warp speeds. The last equation can be used as an approximation for speeds greater than Warp 2. At sub-light speeds, g(f) equals a(v), but the possibilities of another factor could exist for better accuracy. Figure 1d illustrates how a spacecraft moves to an acceleration of 8 a's where one (a) is equal to an acceleration of 32 feet per second per second and the human body experiences a g-force of 8 g's, instantaneously. A g-force of one is what everyone experiences when they fall off a high ledge or 32 feet per second per second. When traveling at sub-light speeds, the values of 8 a's and 8 g's are equivalent to 256 feet per second per second, each. The acceleration speed can be any value that the human body will tolerate. The value of 8 a's was chosen in the earlier chapters. When the spacecraft reaches Warp 1, the vehicle acceleration, a(v), is maintained at 8 a's and the g-force starts to diminish. At speeds greater than Warp 1, the g-force becomes a non-linear function with an exponential characteristic. The curve in figure 1d illustrates that the g-force shrinks or reduces as the warp speed increases and the acceleration of the spacecraft continues to accelerate at 8 a's.

At warp speeds between Warp 7 and Warp 10 a one hundred and fifty-pound (150) man would weigh between one hundred and seventy-three pounds (173) and one hundred and twenty-one pounds (121). Confinement in a special harness would no longer be necessary to subvert the effects of huge g-forces. At any speed above Warp 5, the astronaut can make either of two decisions. He can accelerate his vehicle from 8 a's to a much higher value and reach his destination sooner or he can reverse the acceleration provided by the neutrino trap and start to decelerate. If he makes the latter choice, the process of accelerating, decelerating, and then accelerating again would undoubtedly be assigned to a control circuit that automatically makes the timely choices for him. With a refined control circuit, his comfort in having a body weight of one hundred-fifty pounds (150) would be closely maintained while traveling between Warp 7 and 10. If the astronaut is traveling at Warp 5 and chooses to accelerate more, his body weight will increase above 245 pounds. At Warp 5 he's already experiencing a g-force of 1.6, about the same as taking a curve on a roller coaster, or experiencing the dip in the track at its lowest point. If the astronaut should decide to stop the acceleration or deceleration, the spacecraft will cruse at a constant velocity and the astronaut will experience total weightlessness.

Accelerating at speeds beyond Warp 10, presents a new set of problems; namely, a slow reduction in body weight that will eventually approach weightlessness. Our present day astronauts can relate to the zero-gravity problems with trying to sleep, trying to eat, and trying to poop. Other sets of problems include the need to communicate with the home base or Earth. The next section will address these communication problems.

Additional problems occur when deceleration time is considered. As per earlier calculations, traveling from one warp integer to another whether accelerating or decelerating requires a 44-day wait period if the acceleration or deceleration of the vehicle is 8 a's. Acceleration with the neutrino trap at full sailing conditions is easy to picture. A sailboat does that while sailing down wind. The trick is to sail down wind while the brakes are on! It's difficult to picture a spacecraft slowing down or deceleration in outer space when there are no brakes or friction pads to use. So how is this achieved after stating in earlier chapters that outer space in imaginary time was a frictionless environment? Ideally, our space travelers would want to slow down from Wary 10 to Warp 9 in a 44-day period at a deceleration speed of 8 a's. So how is this accomplished?

Once again we have to rely upon the characteristics of the time shifter. On page 35 in chapter 3 we stated the a spacecraft would pass through Earth's atmosphere at 0.5 parsecs, Earth's oceans at 1.0 parsecs, and through the entire Earth and other heavenly bodies at 3.0 parsecs. Formulas used to calculate the force of friction, as the spacecraft transitions from an environment with friction to one with no friction, are unavailable. However, it is possible to assume the existence of analogue values from one extreme to the other. Consequently, we can make a similar assumption that the spacecraft will decelerate by choosing the appropriate parsec value.

The other method of slowing down is to use the neutrinos emanating from another star. When space travelers have traveled beyond the far influences of our sun, they can sail into a wormhole that leads directly into the black hole at the center of our galaxy, or they can use the strength of the neutrinos coming from another sun. The analogy is a pinball that starts at the top of the game-machine and negotiates its way from left to right bouncing off each bumper without ending at the bottom where the effects of gravity end. Navigation in outer space has got to be the most difficult task when there are no predefined routes. This concept confirms what was said in chapter 2 on page 33; that is, "Earth is just a stopping point on a superhighway." Once a trip is made past three or four solar systems, a trade route is established. Earth just happens to be in the middle of those trade routes!

Communication

String theory is an attempt by man to understand something that his inner voice is trying to tell him! Serious students that try to understand the theory end up frustrated because it goes nowhere. It's unrelated to anything tangible and yet, the inner voice keeps telling us that string theory is a necessity.

If it isn't a necessity right now, then perhaps it's a necessity in the future. Let's ask ourselves, "What resembles string theory that we'll need in the future?". Traveling throughout the galaxy will require two technical skills; namely, navigation and communication. We'll show how string theory is related to communication. Within a nutshell, let's state what we know about string theory, first. String theory is an attempt to "unify gravity with other forces". In his book titled "A Briefer History of Time", Mr. Stephen Hawking states:

[String theory was originally invented in the late 1960's in an attempt to find a theory to describe the strong force (that existed in the cosmos). The idea was that particles such as the proton and the neutron could be regarded as waves on a string. The strong forces between the particles would correspond to pieces of string that went between other bits of string, as in a spider web. For this theory to give the observed value of the strong force between particles, the strings had to be like rubber bands with a pull of about ten tons.]

[Author's comment: In my mind I like to substitute the word neutrino for neutron because I believe that there are more neutrinos influencing our galaxy.]

When I was twelve, I used to climb on top of the roof of my father's house and play with the telephone line that was stretched about two hundred feet between the telephone pole and the house. We lived in a little cottage in Wisconsin with an abundance of trees surrounding our home. The path that the telephone line followed had been cleared of oak, maple, elm, and birch trees. As a kid, I was always investigating and experimenting. So, it was very natural to climb up on the roof and give the telephone line a shake with my bare hand and see what happened. What I saw, was beautiful and fascinating. The telephone line oscillated with one sinusoidal wave that physically traveled two hundred feet toward the telephone pole and then bounced back.

From my observations, I concluded that the speed of the oscillation toward the pole and back was always the same. The only thing that I could control was the amplitude, depending upon how hard I jerked the telephone line. Let's evaluate this wave first, and then we'll be able to correlate communication requirements in outer space. The wave looked like this with the following vector coordinates:

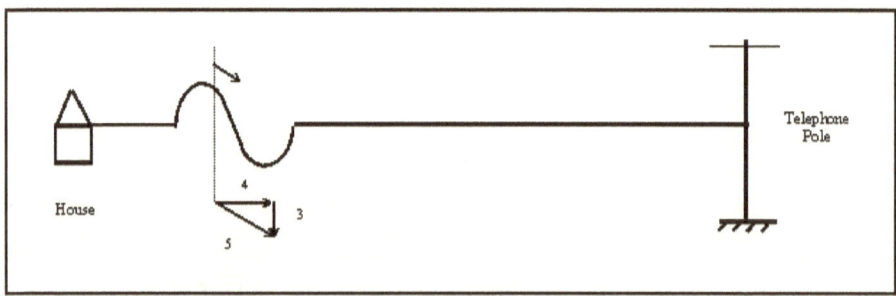

If we use random numbers again to make a point, we can say that the wave was moving at a speed of four (4) miles per hour, and that the line was jerked with a downward force of three (3) miles per hour. The maximum vector speed of the wave is five (5) miles per hour. We'll return to this concept with the understanding that the vector velocity is greater than either the horizontal or vertical velocity.

In the engineering world different board materials have different dielectrics that cause the characteristic impedance to be different. The properties of the media determine the characteristic wavelength of the frequency transmission. A similar example involves oceanic waves beating against the coastline in a rhythmic pattern. The sound never slows down or speeds up. If the sound of the waves were recorded on a record revolving at 33 rpm and then played at 78 rpm, they wouldn't sound right. Whether the rhythmic sound increases in frequency or decreases in frequency, the human ear can detect the difference. If all the salt in the oceans were suddenly removed, the rhythm of the waves would change because of the change in density.

Characteristics of the ocean, Earth's crust, atmosphere, and outer space have a property that is unique to shock waves. When the tectonic plates of the Earth shift under the oceans, tsunamis are created and move across the globe at jet speeds. It's hard to imagine a wall of water traveling at 400 to 600 miles per hour. But it does happen. When the 1906 earthquake devastated San Francisco, it was reported that the fault due to the shift in the tectonic plated traveled at a speed of 5800 miles per hour. When a jet plane exceeds the speed of sound and pierces a new atmospheric front, a cone of noise follows the plane with ear-popping sounds. The water, ground, and atmospheric media react to each event with a shock wave that travels at unbelievable speeds. There's always a "cause and effect" situation. It's logical to assume the outer space will react in a similar manner when a spacecraft pierces the envelope at Warp 1. We can assume that a shock wave will travel through the media of outer space at speeds forty, fifty, sixty, or seventy times greater than Warp 1. We'll use a conservative number like Warp 40 to describe the shock wave and then come back to this concept.

When a stone is dropped into a water pond where there is very little atmospheric wind, a series of ripples emanate from the center. Here again, the speed at which they move away from the center is determined by the characteristics of the media. A teenage boy with a strong right arm can skip a smooth, slick river-rock across a pond seven or eight times. I know this is true because I did it when I was a teenager some fifty-eight years ago. The real challenge is to drop one stone into the pond and then skip another rock on top of the series of ripples that are formed. With perfect timing, each ripple will provide impetus to the second rock allowing it to go faster. The vector speed travels faster than the forward speed of the wave. This is especially true when surfboarding down an ocean wave.

When I was in the Navy and stationed in Hawaii, I used to go surfboarding at Waikiki beach almost every weekend. It took practice learning how to accelerate from zero to the same speed as the wave. The thrill was being able to stand up on the board and control the speed of the board. The real excitement began while trying to surf at a speed greater than the forward speed of the wave. The expression "hang ten" means "let's go faster by moving to the front of the board and hanging ten toes over it". An experienced person can go faster or slower by shifting his body weight forward or backward across the length of the board.

[Author's comment: I got so good at it that I was able to surf with my brand-new wife sitting in front of me. Prior to that I used to go body surfing without a surf board at an Eastern beach where a military sign read, "Off Limits To Military Personnel — several people have drowned here".]

Using the same random numbers that were used in the telephone line scenario, the following conditions are the same. The wave is moving toward the coast with a speed of four (4) miles per hour; the downward speed has a velocity of three (3) miles per hour; and the surfer can travel at five (5) miles per hour. The surfer is moving faster than the wave!

In earlier paragraphs we assumed that a shock wave traveling through outer space moves at a rate of Warp 40. The existence of this shock wave can be used to send a message to Earth. This concept has to be feasible because the alternatives aren't that good. Radio links always travel at the speed of light in atmospheric conditions. If a spacecraft is traveling at Warp 5 and it transmits a signal, the signal will travel at Warp 6 in the forward direction, and at Warp 4 in the reverse direction. I do believe the numbers are additive when traveling in imaginary time. Basically, it's a poor way to communicate. A better method is to use a communication probe and timing mechanism. The analogy is similar to the teenage boy dropping one rock into the water and skipping a second rock on top of the ripples.

The communication probe would consist of an unmanned, miniature spacecraft. It would be launched into the reverse direction that the mother ship is headed with software instructions to reduce speed from Warp 5 to sub-light as quickly as possible. When it makes the transition across the envelope of the space media, it creates a shock wave that propagates in an omni-directional pattern.

[Author's comment: You can't nail Jell-O to the wall but you can get it to quiver whether you tap it from the outside, or tap it from the inside. The crossing of a probe from either direction will produce a shock wave.]

As the shock wave moves across the galaxy at Warp 40, it starts to overtake the mother ship. At that moment a radio transmission is made causing the shock wave to become modulated. If the timing is flawless, the modulation will occur when the shock wave has a vector speed of Warp 50. To justify that a vector speed of Warp 50 is possible, we had to relate scenarios that involved a tsunami, sonic boom, telephone line, and surfboard. How ironic it would be if advanced communication of this nature weren't possible.

The only possible problem is that the radio transmission always has to be in a direction that's in front of the mother ship. The radio transmission will never modulate the shock wave that's in back of the mother ship while moving further away.

Detection and timing the proximity of the shock wave is critical to the success of the operation. The media is a stream of neutrinos. For every second there are 500 trillion neutrinos passing through space the size of a fingernail. A shock wave would cause them to bunch up. Over a specific time frame the flow of neutrinos will increase to 750 trillion per second, decrease to 250 trillion per second, and then return to normal flow at a sinusoidal rate. The shock wave, like the telephone line, transmits a sine wave across the galaxy ready to be modulated by a radio link. In a remote sense, this is what string theory is all about. The cosmos are connected with gravity forces that cause a shock wave to transition itself across the galaxy, instantaneously. Mr. Steven Hawking is correct in speculating that string theory is like a spider web. Only a web with string forces that are built-in, can alert the spider. The specific time frame and the warp speed of the shock wave can only be speculated. The feasibility of advanced communication has got to be real. Why? Let's examine some facts.

The government-funded program called SETI has been in operation for a couple of decades. The acronym stands for "Search for Extra-Terrestrial Intelligence". Mr. Frank Drake first started it in 1960. They have scanned the heavens, listening on special frequencies that they believe are appropriate for space communication. The total success of their operation is zero. They have made no contacts. They have spent enormous sums of money with the assumption that "some one may be trying to reach us" or "we may be able to listen to some UFO chatter". If there is UFO chatter, they're not going to hear it *because they have the wrong equipment.* The analogy is like trying to listen to radio transmissions at short wave, radio frequencies of 28 megahertz from Poland or other parts of the world with a combination AM/FM radio. AM frequencies are quickly attenuated, and FM frequencies operate only on line-of-sight, but short wave frequencies bounce between the Earth and the Earth's atmosphere. The correct frequency band has to be selected. In SETI's case, the correct "dimension" has to be selected to operate in.

To overcome this problem, SETI needs to relocate their equipment into imaginary time. If UFO's are traveling the galaxy at Warp 5, 6, 7, or 8 and communicating on a shock wave traveling at Warp 50, then the possibly of radio transmissions all over the galaxy becomes more real. The SETI project needs to reconfigure its antenna and low noise amplifiers inside an imaginary time bubble and tether it to Earth so that the only communication link is a digital light beam between the equipment and the base station. My contention is that they will hear UFO chatter similar the conversations between personnel in a control tower and the pilots taking off or landing at a major airport. What a rush that would be!

Imagine what a conversation between two aliens traveling in our outer atmosphere would be like. After the translation, it would progress like this:

Spacecraft 1: "Command leader, are we going to land on Earth?"
Spacecraft 2: "No. We better not."
Spacecraft 1: "Why not?"

Spacecraft 2: "The last time we landed we gave the Earthlings our germs."

Spacecraft 1: "Yes, but that was the sleeping bug to help us sleep during our long trips."

Spacecraft 2: "Well, unlike us, it has some bad side effects on the Earthlings."

Spacecraft 1: "Commander, don't you think their scientific researchers would have found a cure for that bug by now?"

Spacecraft 2: "No. They haven't even found a cured for the last virus that we accidentally gave them."

Spacecraft 1: "Which one was that?"

Spacecraft 2: "**Ebola!** Both the Zaire and the Sudan strain. All the bacteria and viruses that we are immune to seem to flourish in Africa and nowhere else."

There's a long pause and the space chatter starts up again.

Spacecraft 1: "Commander?"

Spacecraft 2: "What is it now?"

Spacecraft 1: "I'm worried about some of these trigger-happy pilots shooting us down."

Spacecraft 2: "There's always a few wing-nuts within every group of Earthlings. Just make sure that your parsec time doesn't fall below 0.5 and the bullets will go right through. If that does happen, just blast them with the 'photon-light'. Our power plants can generate megawatts of light power. They'll get the message. Just remember that we're not here to interface with them as per 'divine authority'. We just want to get our supply of water for our hydrogen and oxygen needs and then dump the 'honey-bucket' in their ocean where it will get sanitized and stored at deep levels. Then we're out of here."

[Author's comment: The conversation between the two aliens is totally unscientific. It isn't even connected to any element of factual truth. I apologize to the reader for not locating this conversation in the section titled "A short sci-fi story". I just had to do it because:

1) the rest of the book is serious and this conversation between two aliens is comical,

2) so may people have speculated about the intentions of visitors from outer space, and

3) I just wanted to present my views.

My other reasons for filling up these pages are because:

1) writing about real and the probability of remote concepts is very difficult; my thoughts can only extend so far; and

2) I really want to give the readers their monies worth when they bought this book

Allow me to expand upon the words "divine authority". I really believe that all of mankind is protected by guardian angels and that a sign has been given to alert visitors that Earth is a planet where interference is not tolerated. I even believe that the same warning has been subconsciously extended to VIP's in Washington that provide cover-ups. Remember, I said "subconsciously".]

A Short Sci-fi Story

If you purchased this book in the science-fiction section of the bookstore, you were probably expecting an interesting story other than this documentation of observations mixed with some basic formulas. So we will try to give you a science-fiction story. We'll even make it a story about you, the reader.

The year is AD 2200, and you got tired working for a boss that demanded ten hours of work with only eight hours of pay. History records that this work ethics started in the 1960s when engineers were asked to work overtime without any financial remuneration. Ever since you can remember, you were always interested in flying a galactic spacecraft. People said, "There's honor and respect in being a galactic pilot." So you quit your business job, went to flight school, and graduated summa cum laude. Your expertise in landing on your home planet Earth superseded other pilots.

As you approach Earth in what seems like the millionth time, you announce over the public-address system that the passengers should fasten their special harness because they will be landing at g-forces between 8 and 20. The spacecraft is traveling at Warp 5 when you notice Jupiter and realize that you have 8.7 minutes to decelerate to Warp 0.01. The Earth is traveling at 66,000 miles per hour or Warp 0.01. Your job is to land at that speed and then bring the spacecraft to a halt as Earth spins at 972 miles per hour. To land at latitude 13.5 degrees, your speed has to be 1,000 miles per hour times the cosine of 13.5 degrees or 972 miles per hour. To achieve what amounts to a forced landing, you're going to have to break the sound barrier at around 760 miles per hour with the Earth's surface altitude in consideration. Since the spacecraft doesn't have any heat shields, the first objective is to "land in the sky" by passing through the atmospheric layer in imaginary time. The spacecraft is now at seventy thousand feet above Earth's surface. You cut back on the time shifter, and your optical sensor documents the shift in secondary light from a landmark. You know exactly where you are at on the GIT curve and the amount of gravitational pull the craft is experiencing. Your second objective is to set the gravitational pull by sliding along the GIT curve in and out of imaginary time and create a smooth landing. As you reduce altitude, you adjust the neutrino sail so that your speed is slightly greater than 972 miles per hour. You have to overtake the speed of Earth because you're constantly shifting between real time and imaginary time. This is the only choice because your craft doesn't have any wings or a rocket-type engine.

At twenty feet above the surface you flutter the sail and reduce your speed below 760 miles per hour. There's a loud sonic boom that travels to all parts of the countryside. With a few taps on the time shifter, you gracefully land the craft at 100% gravitational pull.

As the passengers leave, you personally thank them for flying South Nazca West Airlines. One of the passengers asks if everything went all right during the landing. You look at him squarely in the face and smile, "We left another sonic boom trail in the ground like the previous ones that were there 2,200 years ago, but other than that, everything went great!"

A Shorter Sci-fi Story

Most science fiction stories are a tongue-in-cheek event that demands the ultimate imagination from the reader. Both of these sci-fi stories serve a purpose and require the reader to learn the full contents of this book before the full effects of these stories can be appreciated. The purpose of these stories is to convey a set of ideas whereas the book tries to document concepts. The author is merely trying to give the reader a story that swings from extreme imagination to a book documented with a realm of possibilities.

You're a time traveler and it's possible to go back to any time period. The first thing that you want, is to visit Moses and learn about his era. Upon your arrival you notice that he's writing a book. After a warm greeting, you look over his shoulder and notice that he's writing about his ancestors and how long they lived. The passage is a historical time line! Everyone from your time period knows that our solar system drifted across the galaxy between sectors where the measurement of time was different in each sector. You try to explain to him that our solar system crossed a barrier or rib. At first he seems to understand, but then he begins to ask more questions about the barrier rib. As you try to clarify the subject, Moses becomes more and more confused. Finally you give up trying to explain that all of mankind came from a barrier rib. You don't give up entirely because you realize that you can simplify your comments. So you tell him what has been recorded throughout history; namely, "Eve came from Adam's rib." End of story.

Back to Documenting the Book

Scientist needs to stop attributing every mark on Earth's crust as a phenomenon caused solely by a meteorite. There are six different ways that Earth becomes scarred:

- Meteorites and/or comets
- Spacecraft crashing after traveling at warp speeds
- Spacecraft crashing while trying to leave Earth
- Spacecraft landing on Earth and leaving a trail
- Voltage potentials between heavenly bodies
- Gamma burst that exit the black hole and are spewed across the galaxy as an energy burst striking Earth. Refer to figure 1c

The fifth (or voltage) topic will be justified in later chapters. The last topic is difficult to justify. There are only two different events that lend credibility. The first is an article that appeared in the October 2000 issue of *Popular Science*.

> The article states, "So it came as a surprise to astronomers when they found it (the black hole) belching bubbles as large as our entire solar system. The bubbles coming out of the disk create tiny blips in the black holes radio signal."

Another related event correlates radio transmissions that the military detected in our outer atmosphere and was later attributed to gamma burst coming from "somewhere." When the cold war took place, scientists were receiving radio signals every two or three times a day that they thought were caused by the Russians. Scientist will eventually associate the "somewhere" with the black hole. If matter goes in, then matter or energy in some form has to come out. Refer to figure 1c Galaxy Images.

This chapter concludes the topic on Galactic Travels at Warp Speeds in Imaginary Time. The rest of the book will speculate about the changes that occurred in our galaxy so that our space travelers are prepared for what greets them, so that our environmentalist have a better understanding of what's happening on planet Earth, and so that we can reinforce our religious beliefs.

Warp Speeds

Travel Speed (Light-Yrs)	Acc'n Time (Days)	Decel'n Time (Days)	Destination (Light-Yrs)	Distance (Miles)	Direct Travel Time (Days)	Total Travel Time (Years)
1	44	44	*1*	5.8657E+12	365	1.2
2	88	88	*1*	5.8657E+12	183	1.0
5	220	220	*1*	5.8657E+12	73	1.4
1	44	44	*5*	2.93285E+13	1825	5.2
2	88	88	*5*	2.93285E+13	913	3.0
5	220	220	*5*	2.93285E+13	365	2.2
1	44	44	*10*	5.8657E+13	3650	10.2
2	88	88	*10*	5.8657E+13	1825	5.5
5	**220**	**220**	**7**	**4.10599E+13**	**511**	**2.6**
5	220	220	*10*	5.8657E+13	730	3.2
10	440	440	*10*	5.8657E+13	365	3.4
10	440	440	*20*	1.17314E+14	730	4.4
100	4400	4400	*100*	5.8657E+14	365	25.1

Travel Speed (Light-Yrs)	Planet Destination	Destination	Distance (Miles)	Direct Travel Time (Minutes)
5	**Jupiter**	0.0001	**484000000**	8.7
5	**Pluto**	0.0006	**3675000000**	65.9
10	**Jupiter**	0.0001	**484000000**	4.3

Notes
1. Variables for 'Travel Speed' and 'Destination' are entered in columns 1 and 4 as Italic numbers above the double lines
2. Variables for 'Travel Speed' and 'Distance' are entered in columns 1 and 5 as underlined numbers below the double lines
3. Acceleration time is bases upon an acceleration rate of 8 g-force and calculations in chapter 2
4. Distance is destination in light years * 186,000 mi/hr * 3600 sec * 24 hr * 365 days
5. Direct travel time is 365 days*Destination/Travel Speed above the double lines
6. Direct travel time is 365 days * 24 hrs * 60 minutes * Destination/Travel Speed below the double lines
7. Total travel time is the sum of acceleration and deceleration and direct travel times.

The 4TH, 5TH and 6TH Dimensions

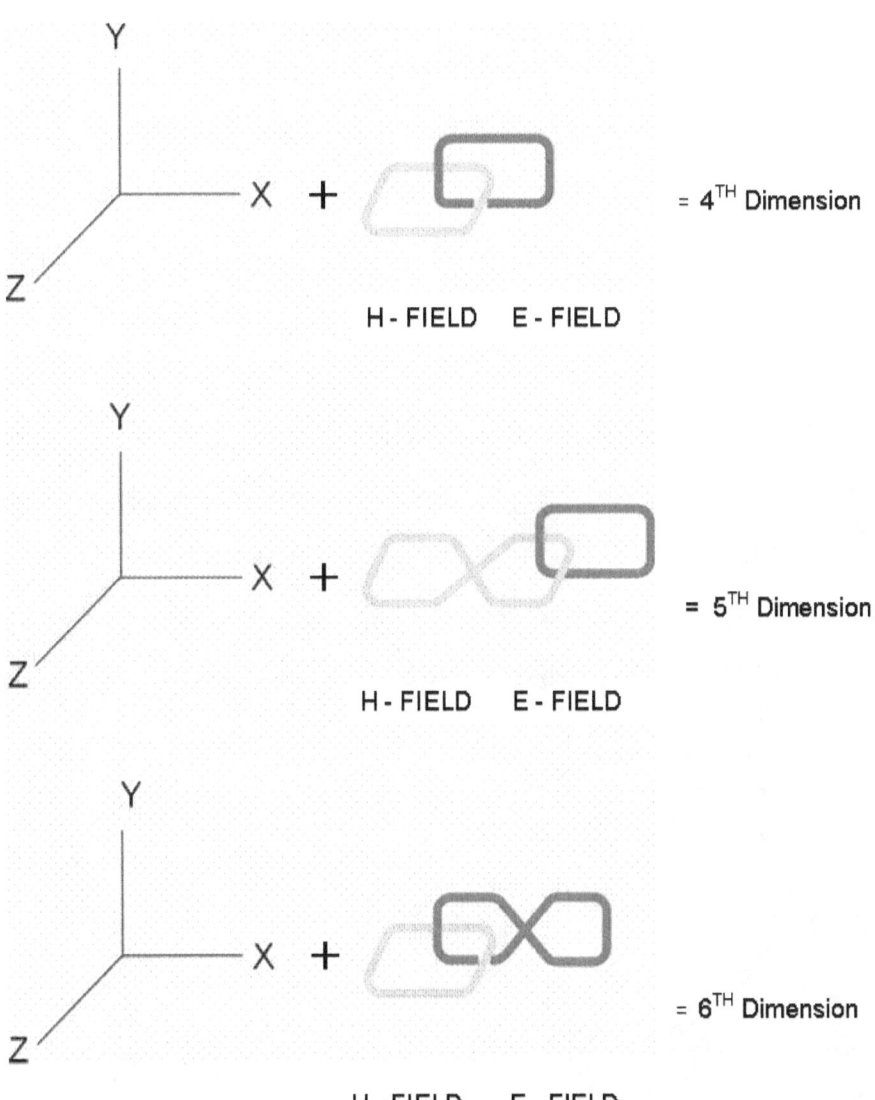

NOTE: Three dimensional space plus electromagnetic wave, in fundamental or harmonic patterns, produce additional dimensions.

FIGURE 1b

Galaxy Image

Source:

Taken from: *http://www.nasa.gov/multimedia/imagegallery/image_feature_996.html*

Information:

NASA's photograph of galaxies, other than our own was presented on their web site to illustrate Black Holes spewing out matter.

Figure 1c

G-force Shrinkage Factor

g-force g(f)	vehicle acceleration a(v)	warp speed W	body weight (lbs) Wgt
8.0	8.0	0.2	1200
8.0	8.0	0.3	1200
8.0	8.0	0.4	1200
8.0	8.0	0.5	1200
8.0	8.0	0.6	1200
8.0	8.0	0.7	1200
8.0	8.0	0.8	1200
8.0	8.0	0.9	1200
8.0	8.0	1.0	1200
4.6	8.0	2.0	693
2.8	8.0	3.0	424
2.1	8.0	4.0	310
1.6	**8.0**	**5.0**	**245**
1.4	8.0	6.0	203
1.2	8.0	7.0	173
1.0	8.0	8.0	151
0.9	8.0	9.0	134
0.8	8.0	10.0	121

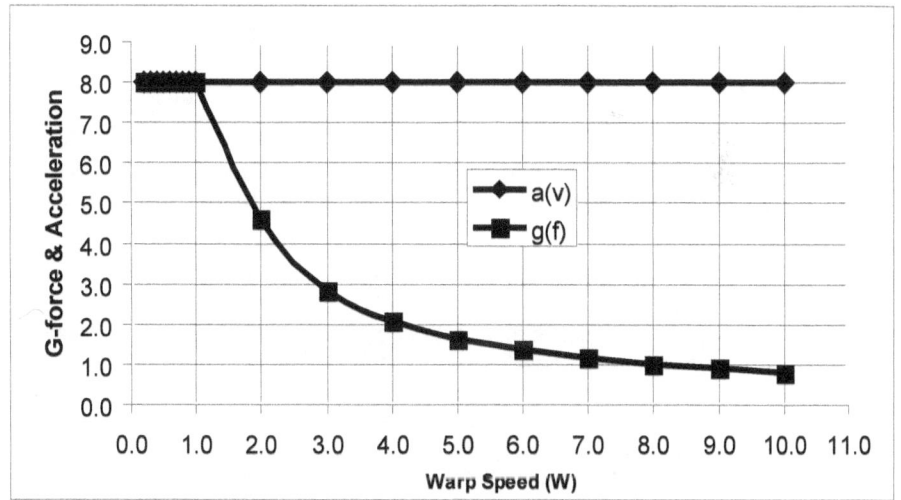

At speeds greater than Warp 1, the g-force is calculated as $g(f) = a(v) / (w^2 -1)^{1/2}$ where w is warp speed. The body weight is based upon a man or woman weighing 150 pounds. A forward vehicle acceleration of one is equal to 32 ft/sec/sec.

Figure 1d

Time Shift Prototype Model

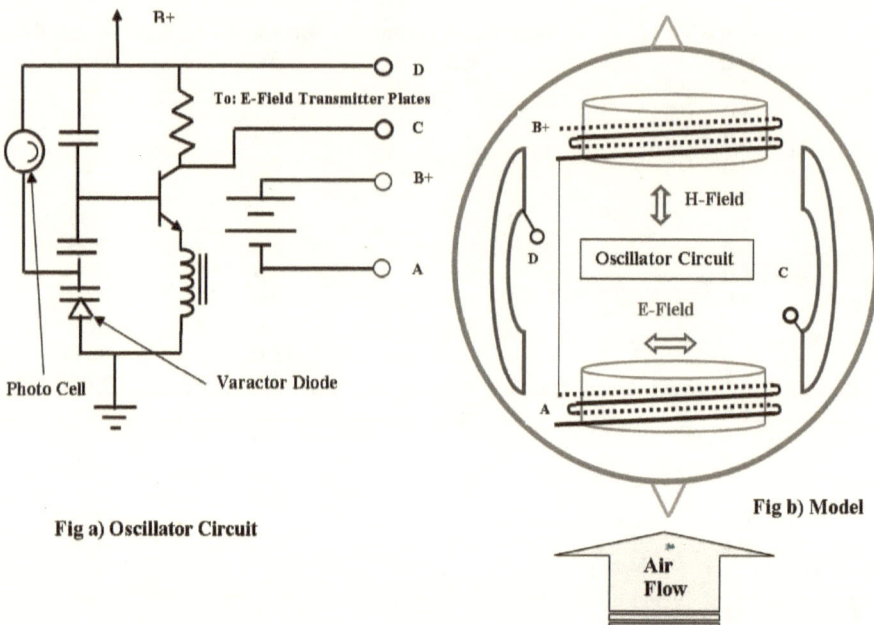

Fig a) Oscillator Circuit

Fig b) Model

Design Criteria:

1. The model is made out of plastic and floats in a stream of air so that nothing touches it.
2. The air stream has to be steady so that the sphere doesn't wobble up and down.
3. The speed of the air across the vanes determines the fundamental frequency of the H-field.
4. The harmonics of the E-Field are controlled by the frequency of the varactor oscillator.
5. The voltage across the varactor comes from the variance of a light beam via the photocell or an RF pick up link. The light beam can be either sinusoidal or pulsed.

The purpose of this prototype is to find the right combination of E-Field and H-Field harmonics that cause the plastic ball to slip into imaginary time. When this event occurs, the ball will disappear. It will stay stationary in space but the Earth will move away. The only connection between real time and imaginary time will be the deteriorating beam of light. Consequently, control of the experiment will diminish as the gravity field is reduced. Refer to the GIT chart or "Light and Gravity vs. Imaginary Time".

Success is achieved when the plastic ball disappears!

Figure 2

Design of the Spacecraft's Time Shifter and Propulsion Engine

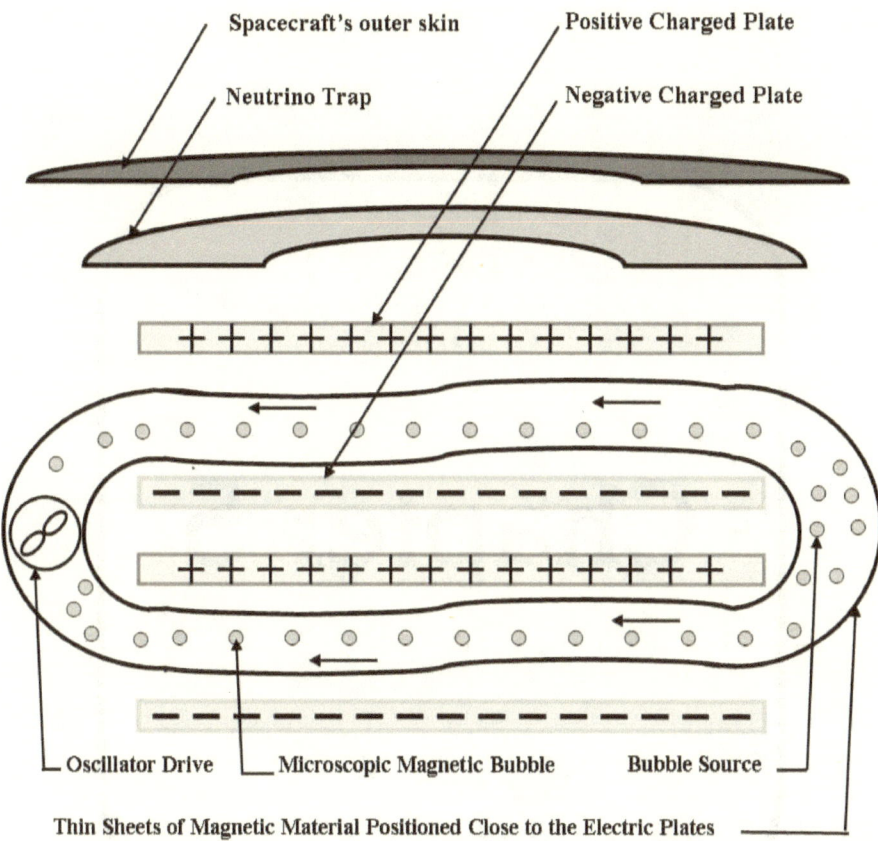

Figure 3

1. The time-shifter is made of a magnetic, metal material approx'ly 1/8 X 1/8 inches. The length stretches across the top of the craft. This material with its magnetic bubbles constitutes the H-Field flowing in the X-plane. The electrostatic field or E-Field is created by positive and negative plates 1/8 X 1/8 inches with charged particles flowing in the Y-plane.

2. Only two "arms" are shown in parallel, but a multitude of arms need to stretch across the top of the spacecraft forming a curved foil similar to the sail on a sailboat. The curvature provides maneuverability. The induction coil that excites the bubbles is not shown.

3. In back of the time-shifter is a neutrino trap and the metal structure of the spacecraft.

4. The neutrino trap is a nanostructure, light in weight that stops neutrinos just as a radiometer of the cover of this book stops photons. It is also a part of the curved foil at the top of the spacecraft.

Figure 3

Chapter 5

MYSTERIES OF THE UNIVERSE

As of this writing, man has been to the moon, scientists have speculated about the possibility of water on Mars, and astronomers have taken wonderful photographs of planets and other galaxies. And yet, nobody knows what to expect when they begin to explore our galaxy. Do we have any clues? Please consider the following hypothesis.

When engineering students are taking advanced calculus and differential equations, they learn that a theorem starts with a hypothesis or tentative assumption followed by an axiom or statement of acceptance as truth. For example, an axiom will state, "A triangle with one ninety-degree angle is a right triangle." This chapter is dedicated to advancing a few hypotheses that lead to theories classified as very unique. These theories can't be classified as theorems because a theorem is a proven fact. Several published books and articles support the theories that are about to be proposed; namely, the Bible, *Worlds in Collision* by Immanuel Velikovsky, and various articles that appeared in *Discover* magazine. When all the information is compiled, the proposed theories become more real. The bottom line is that we're trying to develop a theorem for *sector time*. To understand the merits of Galactic Travel at Warp Speed in Imaginary Time, we need to learn about sector time. We'll start with the hypothesis that sector time is the difference in real time between galactic sectors where a sector is defined as a spiral arm. Figure 4 is a NASA photograph of a messier galaxy M101 taken on March 2, 2006. Our galaxy, the Milky Way, is similar in appearance with spiral arms that are more or less densely packed. The Milky Way has a diameter of about 100,000 light-years and a thickness of about 16,000 light-years. Our solar system lies about 25,000 light-years from the center of the Milky Way and about 8,000 light-years from the edge. Our solar system is very close to the central plane of the Milky Way. There are about 100 billion stars in the Milky Way. Only three galaxies outside the Milky Way are visible with the unaided eye. People in the Northern Hemisphere can see the Andromeda Galaxy, which is about 2 million light-years away. People in the Southern Hemisphere can see the Large Magellanic Cloud, which is about 160,000 light-years from Earth, and the Small Magellanic Cloud, which is about 180,000 light-years away.

Before we start to develop our hypothesis, I have to refer to the facetious story about the scientist who was trying to determine how far a frog could jump when it had lost one of its limbs. The scientist had trained the frog to jump whenever he said, "Jump." Initially, the frog could jump fifteen feet, but with the removal of the forearm, the frog jumped only twelve feet. The scientist wrote these measurements in his journal. Each time another limb was removed, the frog jumped a smaller and smaller distance. Again, the scientist recorded the distance very diligently. After the scientist removed the last limb, he said, "Jump," and the frog just sat there! So the scientist wrote in his journal, "Upon removing all four legs, the frog suddenly goes deaf!"

Unlike this scientist, Velikovsky wrote a remarkable book about the near-catastrophic collision between Earth and its neighboring planets Venus and Mars. The Mynas, Egyptians, Chinese, and the Arizona Indians abundantly document his references with passages from the Bible, the Hebrew Testament, and similar writings. His views are based upon laborious research and reported with subjective observations just as writers did back in 1500 BC, the time of Moses, and in 747 BC, the time of Joshua. If we go back further in time, we can categorize the Great Floor as another cataclysmic event. It's hard to imagine Venus, Earth, and Mars taking elliptical orbits around the sun so close to each other with near misses occurring throughout the ages. But they did. The Earth's orbital path is only slightly elliptical today, and it's 92 million miles from the sun. If all the data for the nine planets and the asteroid belt are plotted on an x-y curve, the information about the slope of the curve begins to tell us that the transition of the orbital paths for Earth and Mars is not smooth.

To achieve a smooth transition, the distance and the axial rotation have to be altered. Ideal results are obtained with Earth and Mars at 124 million and 249 million miles from the sun, respectively. The ideal rotation speed is five hundred fifty and one thousand five hundred days in a year for each planet, respectively. Refer to figure 5a and 5b. Let's speculate that environment conditions were different and that something happened. For more insight, we turn to the Bible.

The Bible was written in code! We all know that. Theologians attempt to decode the Bible with a message obtained from every fifth or eleventh word from a passage. When it's all said and done, they have a word that spells "goobly-goo-goo." The real code is in Genesis 5 and again in Genesis 11 relating Abraham, his sons, their lineage and the sequential path to Moses. The first five books of the Old Testament were written by Moses himself, and there's a high probability that he had help from angels. Human beings are incapable of reciting the longevity ages of their relatives! Those that study theology today tell us that a human being exaggerates when he tries to relate how great a person was or how long they lived. This exaggeration doesn't exist in Genesis 5 and Genesis 11.

> Genesis 5:3 states, "And Adam lived 130 years and begat a son . . . and called him Seth . . . And all the days that Adam lived were 930 and he died. And Seth lived 105 years and begat Enos. And Seth lived after he begat Enos 807 years . . . And all the days of Seth were 912 years, and he died."

Each subsequent passage documents the same longevity ages for eight more descendents, and then we come to Noah. We begin to see an interesting pattern when all the data from the Bible is transferred to three columns titled Begat (years), After (years), and Total (years). Figure 6 contains all the numeric data transferred for the Bible. The italic numbers in columns

1 and 3 and columns 1 and 2 are taken from Genesis 5 and 11, respectively. The math columns are titled Years Lived After First Son, Biological Years, Relative Years, and Time Line. The biological years are calculated by taking Total Years Lived and dividing it by After (years) and then multiplying it by a factor of 1,000. In Adam's case, the results are (930/800)*1000 or 116 years. Why 1,000? I could have used anything close to 1,000, but the factor of 1,000 works out such that Enoch, Methuselah, and Lemech began to live to 122, 124, and 131 biological years, respectively, and then God said in Genesis 6:3, "The lord said, 'His days shall be 120 years'." God's command was fulfilled because the following generations after Noah lived less than 120 biological years with the exception of Nahor who lived 124 years. Consequently, 1,000 is the right number! The biological number of years for each member is consistently within the 107 to 131 range. Refer to figure 6, the Excel chart titled Bible's Time Line Decoded at the end of this chapter.

The real reason the Bible provides a numeric code is so that we can create a time line for the events that happened in the past. In the absence of a calendar, a time line is so important for history's sake. In the column titled Relative Years, the biological years are multiplied by the factor 10. The number 10 is justified because the earliest ancestors lived 10 years longer than the average man of today. For example, Adam, Seth, and Enos lived to 930, 812, and 905 years, respectively. In present time, they could represent three men today living to ages 93.0, 81.2, and 90.5 years. An exact time line is a little difficult to achieve when some of the data is missing in the case of Terah and Abraham. If we choose not to use any part of the data for Terah and Abraham or create estimates for the missing data, then an exact time line can be created. Consequently, only the data in Genesis chapters 5 and 11 will be used. The time line is calculated with each ancestor's time period plus the number of relative years that the previous ancestor lived. For example, Moses's life started around 1500 BC. When Nahor's relative years of 1244 are added to 1500 BC, then Nahor's life started around 2744 BC. Using the same addition for each preceding ancestor, the Great Flood occurred around 11,382 BC, and Adam and Eve started walking the Earth around 22,349 BC. Interesting! The intent here is not to side with the environmental or the creation groups, but doesn't the Bible say in Genesis 2:7, "And the lord God formed man of the dust of the ground and breathed into his nostrils the breath of life, and man became a living soul"? The word "soul" is what makes us different from anyone or anything else whether they lived 1, 2, 3, or 4 million years ago. From this point, we could expand with more theology, but that isn't our goal.

Our goal centers around the factor 10. The factor 10 and the reduction of longevity from Adam at 930 to Moses at 120 years of age indicate that the Bible is trying to deliver a message. Let's assume that the factor 10 is the difference in sector time between two spiral arms in the galaxy. In other words, our solar system could have been in sector A and then moved to sector B.

> The September 2006 issue of *Discover* magazine states, "Instead, we're about two-thirds of the way to the edge [of the galaxy], making a huge loop around the galactic core every 250 million years or so. Astronomers before Shaply had already determined that the sun and its planets are moving toward the constellation Hercules at about 12 miles per second [0.005 % the speed of light] . . . and our main forward speed around the Milky Way at 144 miles per second [0.08 % the speed of light]."
>
> > [Author's note: Some scientists say two-thirds and others say twenty-five thousand light-years, a major difference in astronomy results!]

There are two different clocks between sectors A and B. It's not a difficult concept. A traveler flying from Chicago to London is completely aware of his departure in Chicago time and his arrival in London time and the difference between the two. The real difference between sectors A and sector B of our galaxy is a *magnitude of ten*. The hypothesis is that we lived in sector A many years ago, and that we now live in sector B. If dinosaurs were among us today and they lived to one hundred years, then they would live to one thousand years in sector A and grow big and fat with constant eating. In the *Chicago Tribune*, dated Jan 17, 2008, scientist discovered the skeleton of an eight-foot rat.

> The article states, "The herbivore scurried across wooded tundra of South America 4 million years ago as prey to the saber-toothed cats. Both prey and predator were much larger—a condition that no one has fully explained."

Let's develop a hypothesis that will explain large animals and their association with sector time. When our galaxy was in sector A, time slowed down. The difference in time between the two sectors is attributed to two different (or a combination of) factors; namely,

1) the speed of light is different and/or
2) Earth was at a greater distance from the sun.

Data compiled in figures 5a and 5b illustrate that Earth could have been at a distance of 124 million miles from the sun when our solar system was in sector A. Please refer to figures 5a and 5b again. To achieve a smooth curve for the tangential speed attribute, the author inserted the numbers 124 and 249 million miles for Earth and Mars. If total time was ten times greater in sector A as justified in the decoded Bible, then the length of one day can be calculated as follows:

One day in sector A = 10*(365 days * 24 hours) / (550 days in one year)

= 160 hours

Yes, this explains why animals in prehistoric times got so big! They had eighty hours of daylight to eat and eat and eat!

[Author's comment: These calculations assume the spin of Earth at one thousand miles per hour at the equator has always remained constant. The only way to justify a shorter day is to assume that Earth revolved faster with more kinetic energy. When it moved closer to the sun, it assumed a closer orbit to the sun with a higher energy level. An ice-skater will do the same thing by twirling faster in a circle after pulling their arms closer to their body. The Earth's kinetic energy was transformed into the energy required to exist in an orbit closer to the sun.]

The other explanation for the time difference by a factor of 10 is a little more difficult to grasp. Let's start with something that's familiar. Every science-fiction buff knows that space travelers that leave Earth come back to learn that everyone else has aged. This phenomenon can be demonstrated with Einstein's formula for relativity to show that "time slows as speed of the traveler increases." The time in sector B, where our solar system resides now, has slowed down because the speed of light has increased, and our rate of travel across the galaxy has increased. We are *like* the occupants of a fast-moving spacecraft. If we were to go back to a twin Earth in sector A, everyone would be older. Sector time and the speed of light are related. The speed of light in sector A is slower. It can be speculated the light spectrum had a different hue that caused vegetation to grow more abundant prior to 22,349 BC. This hypothesis on different speeds for light can be broadened to include other possibilities. Let's consider them.

Black Holes

Almost every galaxy in the universe has a black hole in its center. Black holes have an unbelievable gravitational pull on light (photons), neutrinos, planets, stars, and etc. Quasars compared to galaxies emit light more brilliant and intense because they don't have a black hole. Imagine the influence the black hole in our galaxy has on different sectors. Figure 4 illustrates a galaxy named Messier 101 with different sectors that could resemble our own Milky Way. In the center is a black hole that acts like an engine constantly devouring matter from each of its spiral sectors. A sector with more mass contributes more energy to the engine. Using the equation $E = \Delta mc^2$, we can assume that speed of light decreases when more mass is contributed into the black hole. Since there is only one black hole, there is only one engine for the Milky Way. Therefore, another spiral sector that contributes less matter has a speed of light that is greater. Keep in mind, the speed of light equals distance traveled divided by time or $c = s / t$. Equating,

$$E = \Delta m(a) * [c(a)]^2 = \Delta m(b) * [c(b)]^2$$
$$E = \Delta m(a) * (s / ta)^2 = \Delta m(b) * (s / tb)^2$$
where $m(a)$ = mass in sector A and ta = time in sector A and
where $m(b)$ = mass in sector B and tb = time in sector B.

The assumption is made that the black hole is an engine that decomposes matter from all sectors in the galaxy at a uniform rate. With E and s being the only constants in the entire galaxy, then $m(a)/m(b) = (ta/tb)^2$ and the ratio of $ta/tb = (m(a)/m(b))^{1/2}$.

[Author's comment: In a nonmathematical way, I like to think of a black hole similar to a parking meter, one that takes pennies, nickels, and quarters. It consistently gives back a measure of time that is a function of the weight or mass of the coin. Basically, more time at the parking meter is produced with the insertion of a larger coin. This is why E is a constant throughout the galaxy and not c. However, a spacecraft that supplements its neutrino trap with a nuclear engine would have more thrust in sector C than sectors A and B. It could travel farther with the same amount of fuel! Refer to the following table.]

If the total mass of all the stars and planets in sector A is one hundred times greater than sector B, then the clock time in sector A is ten times greater. If the mass in sector A is one hundred times greater, then the speed of light is 1/100 times less than the speed of light in sector B. This is a unique concept because it implies that space travel outside the influence of the Milky Way is more probable. If we define that area as sector C and devoid of any mass, we can see how the speed of light can increase beyond present values. The black hole within our galaxy is an engine. It consumes mass and converts it into sector time with energy (E) being a constant. Without mass to consume, the speed of light in sector C increases as per Einstein's equation $E = \Delta mc^2$ or more precisely, $E = \Delta m*(c|k)^2$ where c|k is the speed of light that is constant for the given sector. Let's create a chart for each sector.

Variables versus Sectors of the Galaxy

Variable	Sector A	Sector B	Sector C
All sector matter	More (100 * m)	Less (m)	Much Less (1/100m)
Speed of Light	Less (c / 100)	More (c)	Much Faster (100*c)
Life Span	930 years	93 years	9.3 years
Biological Years	93	93	93
Time factor	Times 10	Times 1	Times 0.1
Nuclear Energy	0.01 units of measure	1 unit of measure	100 units of measure

Let sector C be defined as an area without stars, planets, moons, meteors, comets, asteroids, neutrinos, and any other matter. These conditions describe an area that exists between galaxies. Areas between sectors of our own galaxy may be devoid of large masses, but they will always have photons and neutrinos within them. Our Milky Way is described as having 100,000-light-year diameter and a 16-million-light-year thickness. The fringe beyond the 16-million-light-year thickness would also be a sector C. If this is truly the case, then why can't space travelers leave sector B and go to back to sector A by way of sector C? Without any neutrinos in sector C, the plan would involve skipping across the sector like a stone cast upon a smooth body of water. Wouldn't this save travel time and allow for greater distances to be traveled? A spacecraft with an booster engine like an nuclear one would provide more thrust in sector C with an energy that is ten thousand times greater as per the equation $E = \Delta m * (100 * C_1)^2$, where C_1 is the speed of light in sector B. If sector C does contain light beams that move one hundred times faster, then is it possible that other galaxies are much closed than astronomers calculate? (I'm drifting.) Let's get back to decoding the Bible.

With time passing much slower in sector A, earlier ancestors in the Bible lived ten times longer. For example, Adam lived to 930 years whereas a life span of 93 years would be believable. At the bottom of the list, Narah lived to 148 years and a live span of 14.8 years wouldn't be believable. If their biological years are calculated with the formula given in figure 6, it can be seen that they *all* lived between 107 and 131 years, and this is believable. In biological years, Enoch, Methuseiah, and Lemech lived 122, 124, and 131 years, respectively. According to Genesis 6:3, "The Lord said, 'His days shall be 120 years'." After that, everyone except Terak lived 120 years or less. We can suspect a typographical error starting with Moses. When Moses was writing Genesis, he got one number out of thirty-six incorrect. By the time Joseph and Moses died at ages 110 and

120 years, the event was over with. What event? The major event that the biblical time line refers to is the physical transition of our solar system migrating from sector A to sector B! Figure 7 titled Rib Theory illustrates orbital paths of Jupiter, Earth, and Mars as they altered from circular to elliptical and back to somewhat circular. (Figure 7 is located at the end of chapter 6 where additional discussions are made.) When planet Earth was in sector A, it was farther from the sun. The number of days in a year and the number of hours were different than they are today. Longer growing seasons under ideal conditions produced more vegetation. Prior to 14,000 BC, the Sahara desert in Northern Africa was fertile and lush with greenery. When Earth moved to sector B, the vegetation died out and turned into desert areas like the Sahara and the state of Montana where the terrain has been referred to as a Serengeti for dinosaurs. The boundary between sectors is essentially a tough barrier to cross because of the differences in time and the speed of light. It's like a protective eggshell surrounding a "yoke and egg." Nothing gets out and nothing gets in. But the yoke can be removed without breaking the shell. If a small pinhole is made at each end of the egg, the yoke and egg can be blown out with air pressure. Like the yoke and egg, our solar system went through the same transition as it crossed from sector A to sector B. The transition took about twenty-two thousand years and got worse during the time of Noah and again during the time of Moses. The crossover occurred in the area that astrologers describe as the Oort cloud. This is an area where more that ten comets have originated and circled our sun. For additional reading, *Discover* magazine presented an interesting article regarding the Oort cloud and its location.

The second hypothesis is that our solar system was struck by a shower of meteors 65 million years ago and started moving toward the boundary of sector B. A speed of 12 miles per second was mentioned earlier in this chapter. Some of the larger meteors struck Earth and eliminated thousand of species, including the dinosaur. Our solar system is like a set of springs. A push on a planet by a meteor/comet is the same as a push on the sun. Even though the rate of travel was very slow, our solar system reached the boundary and started to cross over. Let's assume that our solar system was halfway through the barrier in 8000 BC or ten thousand years ago. Then, if we multiply ten thousand years by 12 miles per second by 3,600 seconds in an hour, by 24 hours in a day, by 365 days in a year, we get 3,780,000 million miles. In other words, the opening in the barrier (or Oort cloud) is 3,780,000 million miles from our sun about one thousand times farther than our distant planet Pluto at 3,675 million miles. With one light-year equal to 5,800,000 million miles, the barrier is slightly less than one light-year away. Our solar system may be traveling at 12 miles per second now that it is in sector B, but when it was in sector A, it was traveling at 12/100 miles per second. We have shown that time and speed in sector A change by a factor of 10 and 1/100, respectively. If we assume that our solar system has been under the influence of sector A for the last three thousand five hundred years or since the time of Moses, then the Oort cloud is only 13,200 million miles away using the same formula and dividing by 100. This smaller number is believable because Pluto passed through the barrier a little over eighty years ago when it was first discovered. Now a new planet called Xena is coming into view at 10,000 million miles from the sun. In conclusion, the Oort cloud is at a distance somewhere between 13,200 and 3,780,000 million miles from our sun.

The third hypothesis is that Mars was filled with vast amounts of water and lost all of it into outer space and here on Earth. Between 22,000 BC and 14,000 BC, Mars took on an elliptical orbit that caused the water to slosh around carving out huge canals, bigger than our Grand Canyon. When Mars got too close to Earth's orbit, it discharged most of the water that appeared on Earth in the form of a flood around 11,382 BC, a period during Noah's time. Other floods are reported in Velikovsky's book *Worlds in Collision*. His book refers to a period around 2300 BC when Emperor Yahou lived and a Chinese book called *The Shoo-king* was written.

The translation that Velikovsky offers on pages 102 to 104 says, "The water was up on the high mountains and the foothills could not be seen at all. The floodwaters prevailed in China for nine years."

Psalms 104 and 107 make reference to water that covered the mountains. It is impossible to determine when the inundation occurred because the psalms do not reference a calendar or a time line. The first written languages, which appeared around 3200 BC, were Egyptian and Sumerian. History began when people started writing about events. Every period before that time is prehistory. The only exception is the story of Noah. The book of Exodus written by Moses and the story of Noah take us back further in time. In conclusion, there were many periods when water inundated the globe but only two were major events. The first one in 11,382 BC was a major one. The second event that most civilizations reported somewhere between 2300 BC abd 1500 BC was cataclysmic but not as damaging as the first. The major flood during the time of Noah was fifteen cubits high (Genesis 7:20). The volume of water that covered Earth is unbelievable when a cubit is converted into eighteen inches and the entire surface area is considered. Today scientists speculate that a natural damn between the Mediterranean and the Black Sea broke flooding the residents living in the basin of the Black Sea. This is highly improbable when a massive volume of water is considered by multiplying all of Earth's surface area by fifteen cubits. A more likely consideration is offered by the biblical time line and the hypothesis of our solar system moving across sectors in the galaxy. These two axioms lead to Rib theory and the feasibility of elliptical orbits of Earth and Mars that produced huge downpours of water from Mars.

Prior to 11,382 BC, land bridges and visible island chains existed between Russia and Alaska, and between Malaysia, Indonesia, and Australia. People migrated whenever weather conditions were favorable as a result of Earth's elliptical path that caused the last Great Ice Age to advance and retreat in repeating cycles. Within the near future, it is conceivable that a research team will validate the correlation between prehistoric weather conditions and Earth's elliptical orbit.

The next chapter associates the annual orbit of Jupiter with the biblical flooding that occurred during the time of Noah in prehistoric times. During historic times, it relates the writings of Moses and the flooding in China around 2300 BC.

This third hypothesis is that Mars had an abundant supply of water; namely, oceans of water that covered most of the terrain. Where is the rest of the water that didn't get dumped on Earth as Mars made its return trips on its elliptical orbit? Simply put, the water is in the form of ice next to the Oort cloud where comets attract the ice with their gravitation field and then burn it off as they circle the sun.

[Author's comment: And then the frog suddenly went deaf! Or are you still with me? If we're going too fast, I can slow down. Seriously, let's proceed with some axioms in the next two chapters.]

Velikovsky points out that Venus had an erratic orbit and came close to Earth with the appearance of a bull with a tail (or contrail) and two horns during the time of Moses. The atmosphere was filled with carbon compounds that percolated down to Earth in the morning in the form of manna (ambrosia) that people ate for sustenance as they were crossing the Sinai Peninsula looking for the land of milk and honey. The manna tasted sweet, and it turned milky when it fell into water; henceforth, the name milk and honey. The appearance of a bull in the sky terrified the people so much that they made an idol in the form of a golden calf that they could pray to. By the grace of God, Moses put an end to idol worshipping; but the people of India, thousand of miles away, saw the same appearance of a bull in the sky and to this day revere the sacred cow as per Velikovsky.

NASA Photograph of Messier Galaxy Dated 2006 March 2

Shown with Sector Arms A & B and Barrier (green)

Source: *http://antwrp.gsfc.nasa.gov/apod/image/0603/m101_hst_f52.jpg*

Information:

This ultraviolet image of the giant spiral galaxy Messier 101 (M101) was obtained by the Ultraviolet Imaging Telescope during the Astro-2 mission of the Space Shuttle Endeavour.

Figure 4

Figure 5a Planet Attribute Chart

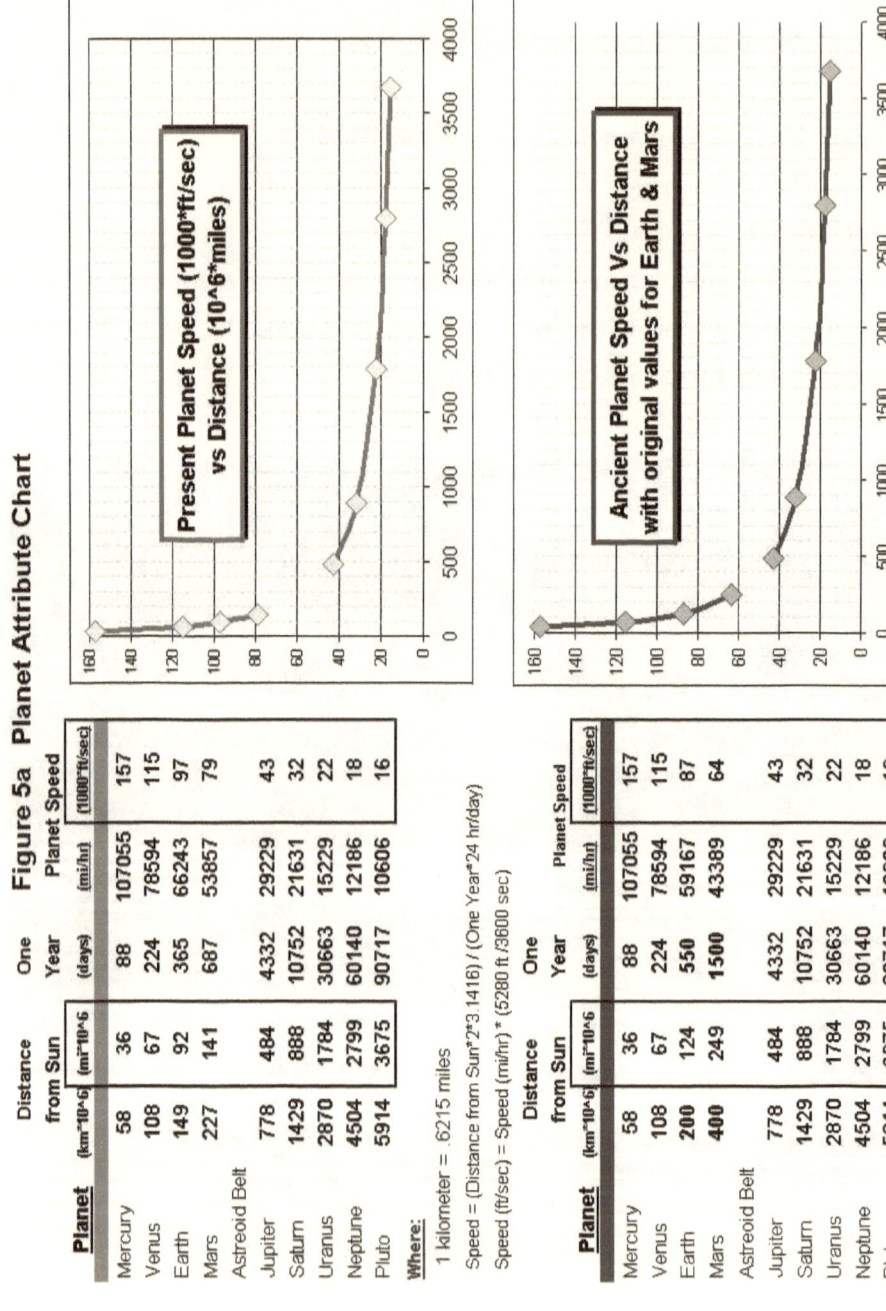

Planet	Distance from Sun (km*10^6)	(mi*10^6)	One Year (days)	Planet Speed (mi/hr)	(1000*ft/sec)
Mercury	58	36	88	107055	157
Venus	108	67	224	78594	115
Earth	149	92	365	66243	97
Mars	227	141	687	53857	79
Astreoid Belt					
Jupiter	778	484	4332	29229	43
Saturn	1429	888	10752	21631	32
Uranus	2870	1784	30663	15229	22
Neptune	4504	2799	60140	12186	18
Pluto	5914	3675	90717	10606	16

Where:

1 kilometer = .6215 miles

Speed = (Distance from Sun*2*3.1416) / (One Year*24 hr/day)

Speed (ft/sec) = Speed (mi/hr) * (5280 ft /3600 sec)

Planet	Distance from Sun (km*10^6)	(mi*10^6)	One Year (days)	Planet Speed (mi/hr)	(1000*ft/sec)
Mercury	58	36	88	107055	157
Venus	108	67	224	78594	115
Earth	200	124	550	59167	87
Mars	400	249	1500	43389	64
Astreoid Belt					
Jupiter	778	484	4332	29229	43
Saturn	1429	888	10752	21631	32
Uranus	2870	1784	30663	15229	22
Neptune	4504	2799	60140	12186	18
Pluto	5914	3675	90717	10606	16

Present Planet Speed (1000*ft/sec) vs Distance (10^6*miles)

Ancient Planet Speed Vs Distance with original values for Earth & Mars

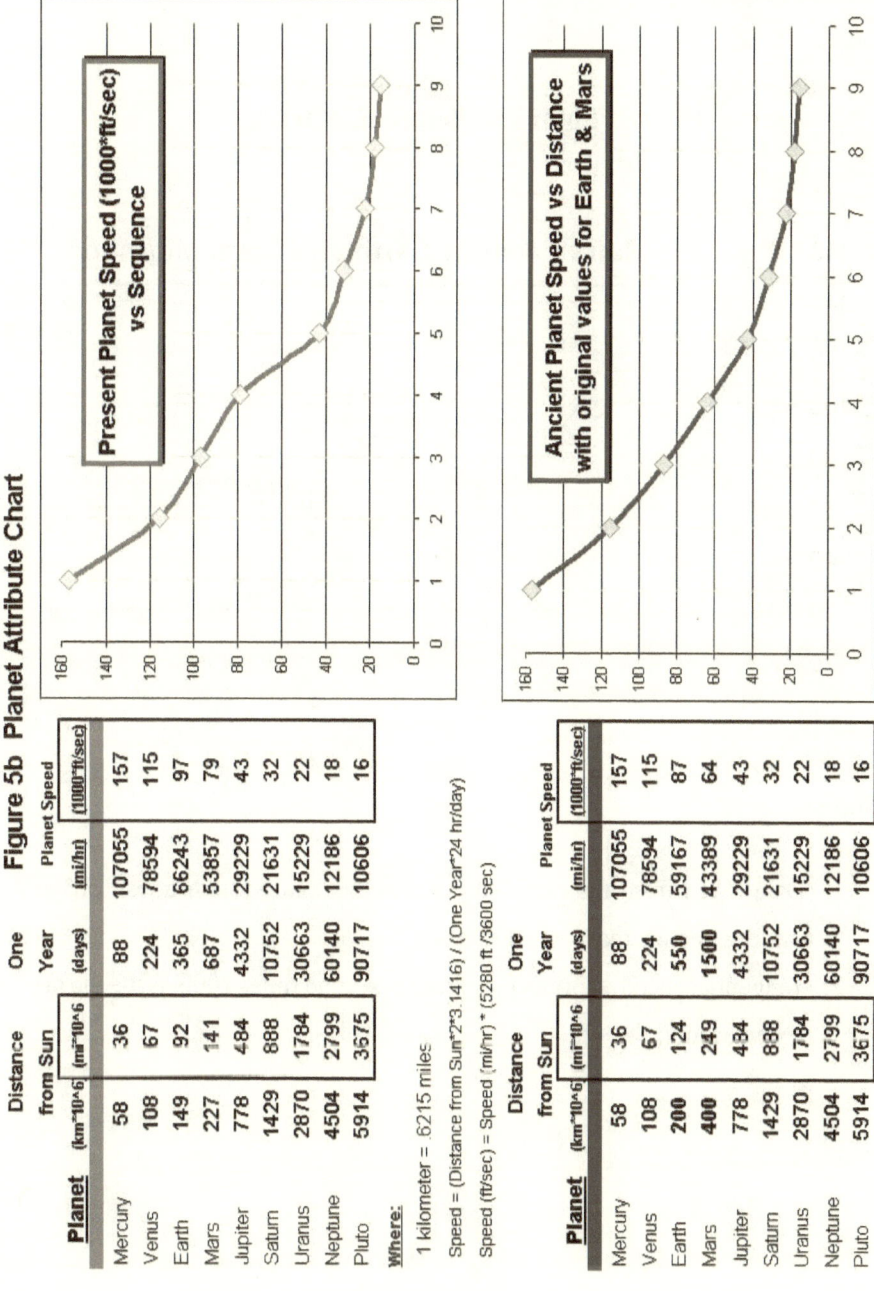

Figure 5b Planet Attribute Chart

Present Planet Speed (1000*ft/sec) vs Sequence

Ancient Planet Speed vs Distance with original values for Earth & Mars

Planet	Distance from Sun (km*10^6)	(mi*10^6)	One Year (days)	Planet Speed (mi/hr)	(1000*ft/sec)
Mercury	58	36	88	107055	157
Venus	108	67	224	78594	115
Earth	149	92	365	66243	97
Mars	227	141	687	53857	79
Jupiter	778	484	4332	29229	43
Saturn	1429	888	10752	21631	32
Uranus	2870	1784	30663	15229	22
Neptune	4504	2799	60140	12186	18
Pluto	5914	3675	90717	10606	16

Where:

1 kilometer = .6215 miles

Speed = (Distance from Sun*2*3.1416) / (One Year*24 hr/day)

Speed (ft/sec) = Speed (mi/hr) * (5280 ft /3600 sec)

Planet	Distance from Sun (km*10^6)	(mi*10^6)	One Year (days)	Planet Speed (mi/hr)	(1000*ft/sec)
Mercury	58	36	88	107055	157
Venus	108	67	224	78594	115
Earth	200	124	550	59167	87
Mars	400	249	1500	43389	64
Jupiter	778	434	4332	29229	43
Saturn	1429	838	10752	21631	32
Uranus	2870	1784	30663	15229	22
Neptune	4504	2799	60140	12186	18
Pluto	5914	3675	90717	10606	16

Figure 5c Before & After

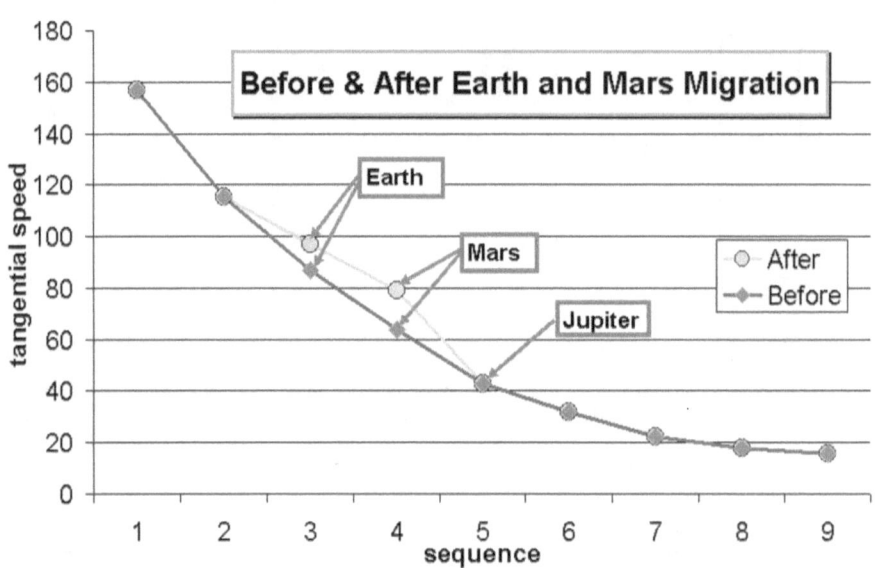

Notes:

1. Data taken from figure 5b shows that the solar system doesn't have the same sequence that it used to have.

2. Tangential speed, expressed as feet per second times 1000, is the the orbital speed around the sun.

Figure 6 The Bible's Time Line, Decoded

Name	Begat(yrs)	After(yrs)	Total(yrs)	Bio-years	Relative-yrs	Time Line(BC)
Adam&Eve	*130*	800	*930*	116	1163	**-22349**
Seth	*105*	807	*912*	113	1130	-21186
Enos	*90*	815	*905*	111	1110	-20056
Cainan	*70*	840	*910*	108	1083	-18946
Mahalaleel	*65*	830	*895*	108	1078	-17862
Jared	*162*	800	*962*	120	1203	-16784
Enoch	*65*	300	*365*	<u>122</u>	1217	-15581
Methuselah	*187*	782	*969*	<u>124</u>	1239	-14365
Lemech	*182*	595	*777*	<u>131</u>	1306	-13126
(Genesis 6:3 "The Lord said, '...his days shall be 120 years'")						
Noah	(150 estimated)	800	*950*	119	1188	-11820
Flood/Noah @ 600 yrs old Genesis 10:28						**-11382**
Shem	*100*	*500*	600	120	1200	-10632
Arphax	*35*	*403*	438	109	1087	-9432
Salah	*30*	*403*	433	107	1074	-8345
Eber	*34*	*430*	464	108	1079	-7271
Peleg	*30*	*209*	239	114	1144	-6192
Reu	*32*	*207*	239	115	1155	-5048
Serug	*30*	*200*	230	115	1150	-3894
Nahor	*29*	*119*	148	124	1244	-2744
Terah	*70*		*205*			
Abra(m)-ham	*86*		*175*			
1.Ishmael						
2.Isaac						
*Jacob(Israel)						
*Joseph			*110*	110		
Moses			*120*	120		**-1500**

Notes:

1. The numbers in italic were taken from the Bible
2. The underlined numbers illustrate how three people were living longer than 120 years.
3. The number 119 for Nahor is possibly a copying error. Everyone else lived 120 bio-years or less.
4. The column labeled 'After' is the number of years they lived after the birth of their first born.

Calculations:

1. Column labeled "Biological-years" is equal to 1000 * column "Total(yrs)" / "After(yrs)"
2. Column labled "Relative-yrs" in equal to 10 * column "Biological-yrs"
3. Column labled "Time-Line" is the accumulation of all "Relative-years" starting with Moses's birth.

Chapter 6

AXIOM

In the mathematical department, axioms are statements accepted as true and used to support an argument or inference. To develop a theorem or proof, the following progressions of stages are used in this book: (1) hypothesis or proposal, (2) axiom, (3) theory, and (4) theorem. This chapter will cover axioms that support Rib theory, dark matter, Mayan predictions, global warming, oceans on Mars, Greek mythology, visitors from space, and the Bible. The introduction of Rib Theory in the previous chapter has a multitude of interesting ramifications.

Rib Theory and Dark Matter

Figure 7 illustrates the path that the solar system took as it passed through a barrier or rib. These barriers surround each sector in the galaxy and keep the circular momentum of the galaxy uniform. According to astronomers, our galaxy has made eight revolutions in the last billion years. They are confounded as to why the galaxy doesn't wind down on itself like a batter of pancake mix that will move faster when it's closer to the center. In other words, why do the stars at the outer fringes have the same tangential speed as the stars closer to the center? In reality, the circular spin of the galaxy is similar to specks of dust spinning on a record player. Some kind of force that can't be seen holds them so that they have the same tangential speed. That force is called a rib. What is it made out of? Well, what is the wake that trails a speedboat made out of? The wake is made out of water and air bubbles just as a rib is made out of more matter in the galaxy, and probably a stream of neutrinos flowing toward the black hole along the aisle or paths at the fringes of each sector. Scientists don't know how to explain a solar system that doesn't wind down on itself, so they attribute it to dark matter and dark energy. In reality, it is the ribs that keep the stars circulating the galaxy like dust particles on a record, turning on a record player. The May 2001 issue of *Discover* magazine has an article on page 19 titled "Ripples in Space." To quote:

> "Recent evidence suggests that as much as 95 percent of matter in the universe consist of some exotic substance that neither emits, reflects, nor absorbs any kind of light . . . . gravity waves might help map this so-called dark matter for the first time."

The December 2001 issue of *Discover* magazine has an article on page 16 titled "X-Ray Eye on Dark Matter." To quote:

> "The map shows that dark matter is more common of the cluster (in the constellation Draco), and it offers the best look yet how dark particles clump together."

Now we know what ribs are! Ribs give our Milky Way the strength it needs to hold its shape, and the strength that it needs to keep its shape. The drawing in figure 7 shows our solar system passing through a rib that is so strong only the massiveness of Jupiter can keep it open. After Jupiter passed through, Mars and Earth have difficulty and assume elliptical orbits. The ice ages prior to 10,000 BC could probably be explained by Earth's elliptical orbit when it was farther from the sun. Prior to 10,000 BC, every winter averaged colder weather with more glacier buildup. After 10,000 BC, the ice age ended as Earth's orbital path began to decay closer to the sun. With the exception of a mini-ice age that occurred around AD 1400, Earth's atmosphere has been warming up. The two elliptical orbits of Earth and Mars lasted almost ten thousand years until today Earth has a circular path that is only slightly elliptical and Mars is still making the transition.

[Author's comment: Back in the late '90s, I remember news broadcast that Mars and Earth would be at their closest proximity. It was announced that a newspaper could be read at night by the light that Mars reflected. I tried it, and it was true.]

Figure 5a was introduced in the previous chapter. It illustrates the attributes of the planets; namely, tangential speed versus distance from the sun. The curve in the upper right corner demonstrates that the transition of a smooth line doesn't exist. When different numbers are picked for Earth and Mars as indicated in bold, the transition of the curve at the bottom of the page becomes smoother. Figure 5b plots tangential speed of each planet against a numeric sequence. Close examination indicates that calculation of the planet speed with today's data doesn't have a geometric progression. For example, the numbers ¼, ½, 1, 2, 4 have an ideal progression. If we pick numbers like 124 and 249 million miles to represent the distance from the sun for Earth and Mars, then we can see a better geometric progression. The physicist Niels Bohr (1922) postulated that atoms with a multiple number of electrons circulating the nucleus have different orbits or shells by which there is a progression of electrons like 2, 4, 8, 16, etc., for each orbit or shell that exist farther from the nucleus. No one has made an analogy between atomic physic and heavenly bodies, but it is possible that one does exist. Figure 5b gives a pictorial view of a geometric progression showing a before and after. Prior to 22,000 BC, our solar system probably had nine planets and a dozen more with nearly circular orbits formed by the natural growth process of a star and the solidification of planets. Our solar system had a uniform progression of planets from the sun that developed naturally. Figure 5c illustrates how a uniform progression used to exist.

As stated earlier, conditions were different; namely, man and animals lived longer, especially dinosaurs that ate all the time and grew big. The skeletons of wolves, sloths, and predatory saber-toothed cats were much larger than their equivalent species of today. If a beam of light in sector A travels at 1/100 its speed, then perhaps it shines with a different color and provides more nourishment for plants and animals higher on the food chain.

[Author's note: There are places on Earth where the sun shines different. If you go to Tucson, Arizona, and look at the surrounding sunlight, you'll notice the difference from when you get off a plane in San Jose, California, and make the comparison. Wrigley Field in Chicago looks as bright and intense in vivid color as it does on TV.]

Figure 7 illustrates how the planets squeezed through a narrow opening in the barrier where the Oort cloud resides. To this day, the process continues. All nine planets have made it through the opening. (And we're reestablishing Pluto as a planet in this book!) About a dozen or more planets (or objects) take on a figure-eight orbital path around the sun and continue to migrate from sector A to sector B and back again. A new planet called Xena is totally in sector B now. The rest consist of more planets and smaller bodies or comets that pass through the barrier, circle the sun, and return to sector A. Approximately ten comets show up each year.

The May 2006 issue of *Discover* magazine states,

"Nothing in the Oort cloud is visible directly through telescopes, but astronomers infer its existence because it occasionally spits out objects that plunge toward the sun where they sprout long, vaporous tails and become comets . . . Today's best telescopes can penetrate only the nearest part of the solar system's outer regions, known as the Kuiper belt . . . Xena sits at a rakish 45-degree angle to the main planets, making a mockery of the old idea that the Kuiper belt is literally a belt."

The last statement about Xena's forty-five-degree orbital path is further proof of a portal in the barrier that is continually opening, closing, and shifting its axis with respect to the solar system from perpendicular or ninety degrees to forty-five degrees. Figure 7 attempts to illustrate this three-dimensional concept with a limited two-dimensional drawing. The contentions for Rib theory are stated in figure 7 and again as follows:

- Based upon the time line decoded from the Bible
- Based upon Jupiter having an annual rotation of 4,332 days (11.87 Earth years) and being more influential than Saturn, Uranus, Neptune, or Pluto
- Based upon a barrier that surrounds each spiral sector in our galaxy
- Since the barrier doesn't lend itself to the passage of smaller planets, the aperture opening tries to close when Jupiter isn't near.
- Figures 7b, 7d and 7e show the aperture at its maximum width with Jupiter closest to the barrier/rib.
- When the solar system was passing through the barrier and the aperture opening was at its minimum, the following happened:

  a. Mars released its oceans of water to the gravitational pull of Earth when their orbits were elliptical. (Since Earth is more massive than Mars, the gravitational pull caused objects like rocks and water to migrate across space toward Earth.)
  b. Some of the water settled near the Oort cloud where it resides today.

Rib theory also states that Earth and Mars moved closer to the sun by 51 and 173 million miles, respectively. Refer to the planet attribution chart. Figure 7a shows the natural formation of planets in a circular formation with

Earth at a greater distance from the sun. Figure 7c shows the relationship between three planets with Jupiter removed from the barrier that is causing the orbits of Earth and Mars to compress into elliptical ones. Figure 7c also shows Earth and Mars in very close proximity when major catastrophes that occurred in 11,382 BC. Figure 7e shows the start of a reoccurrence of catastrophes between Earth and Mars around 1500 BC. At that time, Mercury and Venus maintained their circular orbits (not shown), but Earth with its elliptical orbit came very close to Venus. The position of the planets and the solar system with respect to the barrier opening are all approximate illustrations.

Mayan Predictions

The Mayan that used to live in the Yucatan Peninsula as far back as 3114 BC predicted a doomsday would occur on December 21, 2012. The Mayan astronomers observed the sun when it passed directly over their temples or Zenial passages and when it rose due east and due west. From the top of their towers at Chichen-Itza, they observe the equinoxes when days and nights are equal and thereby establish a clock for the annual passage of time. With this time clock, they established a calendar that was 5,126 years in length (3114 + 2012). Over the years, they watched the progression of the constellations and correlated it with the crossing of the winter solstice sun with the constellation that is between our solar system and the center of the galaxy where the black hole resides. Figure 1c illustrates a galaxy similar to our own turned on its side with a thickness of sixteen light-years. Our solar system is eight light-years within either edge of the galaxy. All the orbits of the nine planets form an ecliptic plane that is moving above and below the centerline of our galaxy over a thousand year cycle. Astronomers have confirmed that on December 21, 2012, the sun, Earth, the star cluster Pleiades, and black hole will all be in a straight line. The gravitational pull by the black hole is supposed to be the strongest along this straight line. Consequently, the Mayans predicted a doomsday because a solar flare (or CME, coronal mass ejection) could shoot out with the pull of gravity and scorch Earth.

[Author's comment: I don't think there will be a doomsday because of the neutrino effect. If you can believe in Rib theory, then you can understand how neutrinos will divert a solar flare along sector lines. Figure 4 illustrates how a spiral galaxy has sectors that twirl down to the center of the galaxy at an angle something less that ninety degrees. All these neutrinos that come from our sun and other suns behind our sun stream toward the black hole at an angle to the straight line that the Mayans proposed. Rib theory is about a powerful force of neutrinos that can change planetary orbital paths into elliptical ones and divert solar flares.]

Great catastrophes occurred in 11,382 BC and 1500 BC. According to the ancient writers, Venus was associated with catastrophe more so than Mars because the minimal distance between Venus and Earth with its elliptical orbit that was closer than it is today. The start of history is defined as a time when ancient people started writing. According to Marilyn vos Savant's research efforts, her article appeared in the *Chicago Tribune* on January 27, 2008.

"The first written languages which appeared around 3200 BC, were Egyptian and Sumerian. The earliest written languages that still survive are Chinese and Greek. In written form, they date back to about 1500 BC."

Earlier we quoted Velikovsky and his research into the Chinese, Hong Kong edition titled *The Shoo-king* that made reference to a flood in

2300 BC. The inference here is that early man in prehistoric times conveyed stories from generation to generation.

> [Author's comment: My grandfather was born in Sicily where he learned of a legend that he believed true. When I was twelve years old, he told me about the existence of fast-flowing water stream like a jet stream that flowed under the Mediterranean from Greece to Sicily. He talked about people who would communicate between the two countries by dropping a bottle with a message inside the whirlpool at each end. I really believe my grandfather for about ten years. When I was a young man and started reading Greek mythology, I discovered a similar story. My grandfather was passing down a myth that I have remembered for fifty-eight years!]

In conclusion, there won't be a doomsday in 2012. The Mayans were good astronomers, but they got it wrong. They didn't know about the Bible's time line, spiral arms, and the passage of our galaxy from sectors A to sector B as per Rib theory.

Global Warming

The present conditions in figure 5a show that Mars and especially Earth are too close to the sun. A short question-and-answer exchange produces the following:

Q: Is this an unstable condition for our solar system?
Answer: Don't know
Q: Is global warming feasible?
Answer: Yes, since the year 22,000 BC, we have moved closer to the sun by 26% or 100*(124-92 million miles) / 124.

Q: Is mankind contributing to global warming?

Answer: Yes. A certain percentage is due to mankind, and the rest is due to Rib effect. The prediction of another ice age base on the strata of ice cores can't be used to forecast optimistic conditions in the future. Reality can't be optimized when it is based upon conditions that existed ten, twenty, thirty, or one hundred thousand years ago. To prevent global warming, countries through the world need to make the following changes now:

1.   Build carbon dioxide collectors and bury the compound under the oceans in liquid form.
2.   Modify existing coal-to-electric plants to accommodate coal scrubbers that remove undesirable impurities.
3.   Encourage new start-up companies to build coal-to-gas-to-electric plants that remove carbon dioxide and bury it under the plant.
4.   Initiate construction of futuristic power plants that utilize neutrino traps with the ability to convert motion into static electricity and powerful currents. [Refer to chapter 4 and the presentation that was made on power plants created from nanostructures. Earlier chapters indicated how the kinetic energy of a neutrino when harnessed, can exceed the photon energy that drives a radiometer. Hopefully, another book will detail the construction of power plants that replace gasoline engines and provide environmental, friendly electric power to heat homes.]

In her book titled *Mythology*, Edith Hamilton relates Ovid's story about Phaethon who was a mortal on his mother's side and learned about by his father, the sun god. He asked permission to drive his father's chariot (the sun) across the sky. His father was bound to a promise that he couldn't break. This story certainly isn't an axiom, but it does have a romantic connotation and possibly, a resemblance of some catastrophe that happened in the past. On page 183, the story continues with Phaethon in trouble:

> The horses soared up to the very top of the sky and then, plunging headlong down, they set the world on fire. The highest mountains were the first to burn. Down their slopes the flames ran to the low-lying valleys and the dark forestlands, until all things everywhere were ablaze. The springs turned into steam: the rivers shrank. Looking down from Olympus they saw that they must act quickly if the world was to be saved. Jove seized his thunderbolt and hurled it at the rash, repentant driver. It struck him dead, shattered the chariot, and made the maddened horses rush down into the sea.

On page 29, the goddess Athena is described as follows:

> She was the daughter of Zeus alone. No mother bore her. Full-grown and in full armor, she sprang from his head.

Oceans on Mars

In Greek mythology, Athena symbolizes the planet Venus, and Zeus (or Jove in Greek) symbolizes the planet Jupiter. As stated earlier, the "horns" on Venus resembles a bull that the followers of Moses began to worship. In his book *Worlds in Collision*, Velikovsky describes a condition of catastrophic proportions with earthquakes, volcanic eruptions, hail combined with fire, and the parting of the Red Sea that could only be caused by the proximity of two other planetary bodies; namely, Venus and Mars.

Figure 7 shows elliptical orbits for Earth and Mars. Correlation exists with Mars causing major flooding in 11,382 BC and again with significant flooding in 2300 BC. The major catastrophes were attributed to

Venus around 1500 BC or the time of Moses. If this hypothesis is correct, then the barrier would have squeezed the orbital path of Jupiter first, Mars second, and Earth third. Venus would have been spared, but to an observer on Earth, it would appear that Venus sprang from Jupiter. Venus didn't come from Jupiter but rather maintained a steady orbit while Earth came to it. We have to assume that the distance between Venus and Earth was so close that it caused the tectonic plates in North Africa to shift. With Earth revolving around the sun in an erratic orbit decaying by 26%, there had to be events when extreme temperatures parched the land and dried up the rivers. When Earth and Mars had erratic orbits, vast amounts of water, stone, and red clay fell down on Earth. History books of all periods support this argument.

> The June 2000 issue of *Discover* magazine states, "The channels, more than 100 miles wide and 1,000 miles long, might have formed when floodwaters poured from the southern highlands into an ocean covering the northern lowlands (of Mars)."

If oceans can be pictured on Mars, then canals are the result of massive waves carving away a canyon five times larger than our Grand Canyon. If water did exist on Mars, then why has our government spent resources and tax payers' money on rocket ships that traveled to Mars to prove that "water could have been there"? Why prove the obvious? And why use rocket ships that eventually will become outmoded? Chapter 1 discussed how a trip to Mars would take only 67 hours traveling at Warp 0.1, plus 10 hours for initiation time, and 8.3 hours for acceleration time.

Visitors from Space

Why did all these catastrophic events occur during the time of Moses? Answer: Because fear promotes religion faster than intellect does.

On a scale of ten, fear of dying is a 10, fear of a calamity is an 8, and fear of being chastised or scolded is about 2. When mankind is told that they did something wrong, the tendency is to shake it off. Intellect has nothing to do with religion, unless the principle is comprehended. It grieves this author that there are so many intelligent beings today that are total atheist. If faith and/or works in faith won't wake them up, then let's appeal to their intellect. The Bible is in code because early man would have never understood the principles. What principles? Answer: Knowing where we came from so that we will understand where we're going.

Erich Von Daniken was the first author to speculate about the peculiar marks in the Earth's surface on the plains of Nazca with his book *Chariots of the Gods* (copyright 1968). In 1995, Graham Hancock revisited the subject with his book *Fingerprints of the Gods*. Both authors wrote great books with good data, great information, and monumental efforts. At the time, their books were somewhat controversial because they touched on a new subject. Von Daniken was labeled as a romantic that favored the visitation of aliens. After visiting the site, Von Daniken discussed the thirty-seven-mile long plain being devoid of primitive surveying tools that could have been used to create the lines.

Both authors agree that the plains of Nazca were the landing ports for alien spacecraft. To quote Hancock on page 40:

> Similar attention to detail is to be found in the geometric devices. Some of these take the form of straight lines *more than five miles long*, marching like Roman roads across desert, dropping into dried-out river beds, surmounting rocky outcrops and never once deviating from true.

Then again on pages 49 and 92, Hancock continues,

> The mystery was deepened by local traditions which state not only that the road system and the sophisticated architectural

had been "ancient in the time of the Incas," but that both "were the work of white, auburn-haired men" who had lived thousand years earlier.

The image I could not get out of my mind was the Viracocha people leaving, "walking on waters" of the Pacific Ocean, or "going miraculously" by sea as so many legends told.

Descriptive words in these two books talk about straight lines that drop into riverbeds and cross over rocky outcroppings. The entire plain is covered with baseball- to football-sized rocks and a surface that changes color when an inch of the surface is removed. With very little rain in that part of South America, the etched conditions remain preserved for thousand of years. Books document the events at Nazca as happening between 500 BC and AD 500. Let's ask ourselves, "What else happened two thousand years ago?" Hold that thought, and we'll come back to this part.

In chapter 4 titled Warp Speeds, the problems of slowing down a spacecraft traveling in imaginary time were discussed. This could be accomplished with helicopter landings at night or forced landings in the daytime. With Earth spinning at one thousand[1] miles per hour and the spacecraft jumping in and out of imaginary time, a sonic boom at low altitudes can change the landscape, separating the rocks into a five or ten foot swath before bring the craft to a halt. The existence of these lines supports the proof of imaginary time!

Now, let's ask ourselves why visitors would want to come to the barren plains of Nazca with nothing more interesting to do but entertain the natives with coloring books like monkeys, spiders, and whales that they helped them etch in the soil, hundreds of yards long. This is the strongest axiom that I have yet to make. The white auburn-haired men

---

[1] The speed of the Earth's rotation at any location is 1,000*cos (latitude).

who "walked on the waters" across the Pacific were visitors who came to witness the birth of Christ. With the same appearance and features as us, they blended with the people of the Holy Land to document the life of Christ. They were the three Kings that really did come from afar, as per the song "We Three Kings of Orient Are," "bearing gifts we traverse afar. Field and fountain, moor and mountain, following yonder star" by a contemporary composer, John Henry Hopkins. Let's ask ourselves, With mankind being what it is today, would any king or ruler back then have the compassion to leave his busy, daily routines and visit the Christ child? Could the president of a country be convinced today of his need to visit an important happening? Let's face it. The seven deadly sins permeate all of our ranks.

The Bible

The Bible refers to a period of time when mankind was more virtuous, a period when Enoch and his followers were closer to God.

Genesis 5:23 and 24: "And all the days of Enoch were three hundred sixty and five years: And Enoch walked with God and he was not; for God took him."

The story of Enoch is so different from all the others who lived between eight or nine hundred years that you have to wonder if the visitors who landed at Nazca were the descendents of Enoch along with his small band of followers.

As a final note to this book, we'll make reference to that thin line of demarcation between religion and science. As more scientific facts are uncovered every day, more people are losing their faith and their belief in perpetual life. Technical advancements are altering our beliefs. We have covered topics about imaginary time, the abundance of neutrinos traveling at warp speeds, sector time, different sectors of our galaxy with ribs separating them, and finally water on Mars. If any of these hypotheses are proven true, it won't shake my faith in God, and it shouldn't change yours but rather bring you closer. Einstein once said that God doesn't play dice. Of course not, he would know the outcome every time! His long-range strategies are beyond the scope of our best minds and the skills of our best pool players and chess masters. Consider this thought. If a meteor hadn't struck Earth 65 million years ago, our solar system would have never started to drift across the galaxy. All the biblical events surrounding Noah and Moses would have never taken place. Dinosaurs would still be roaming Earth devastating mankind. Anyone who can orchestrate all the variables and combination of events so that the "big" lesson is learned has love for his chosen people and all of mankind. Let's think about it. He started on a small scale with Abraham to test his resolve by asking Abraham to submit his son's life as an offering. When mankind needed to learn the lesson again, he worked miracles through Moses reaching out to a larger group. After one thousand five hundred years had passed, mankind still hadn't learned the lesson, so he sent his only son to help an even larger group. In each of the three cases, the group that was supposed to learn the lesson got larger and larger. He can't reach out anymore. We have had three chances to get it right.

> Allow me to quote Saint John as he quotes Jesus, **"Let not your heart be troubled: ye believe in God, believe also in me. In my father's house are many mansions: if it were not so, I would have told you. I go to prepare a place for you."**

I'm going to put on my theologian hat one more time and say that this passage has the most inspiring words about free will in this world and the next. You'll always be able to say and do as you think, a very powerful thought, more powerful than all the technical advances within the last sixty years. I say sixty because I can remember riding in a buckboard drawn by a horse on my way to school that is now in the center of Phoenix, Arizona. I can remember the technical advances starting with an eight-inch TV with a magnifier glass and the first computer that I purchased in 1990.

Figure 1b illustrates the feasibility of different dimensions that are additions to the three-dimensional world we experience every day. We have to assume that they can't be build on each other but rather exist independently without any overlapping. For example, the fourth dimension is for communication; that is, radio waves that travel at light speed. The fifth dimension has multiple number of H-field harmonics that produce a time shift and the ability to accelerate beyond the speed of light. The sixth dimension has a multiple number of E-field harmonics and a speciality for the flow of electrons, ions, or charged particles.

I was going to write several chapters about the sixth dimension and how it relates to the spiritual world, but I'll leave that to your imagination. I could write a few thousand words about this subject, but I'll save that for the next book. However, I will say just a little. Early man believed in the Elysian fields as per mythology myth, and now more advanced thinking brings us to a belief in resurrection by a different means. As reported, when people die, they experience a "light at the end of a tunnel" and a sensation of being whisked along at a *very rapid speed*. More people will believe that this is true when imaginary time is proven to exist. When you die, you take your conscience filled with good deeds and bad deeds to meet your maker! Allow me to say that one more time, "When you die, you take your conscience filled with good deeds and bad deeds to meet your maker."

RIB THEORY

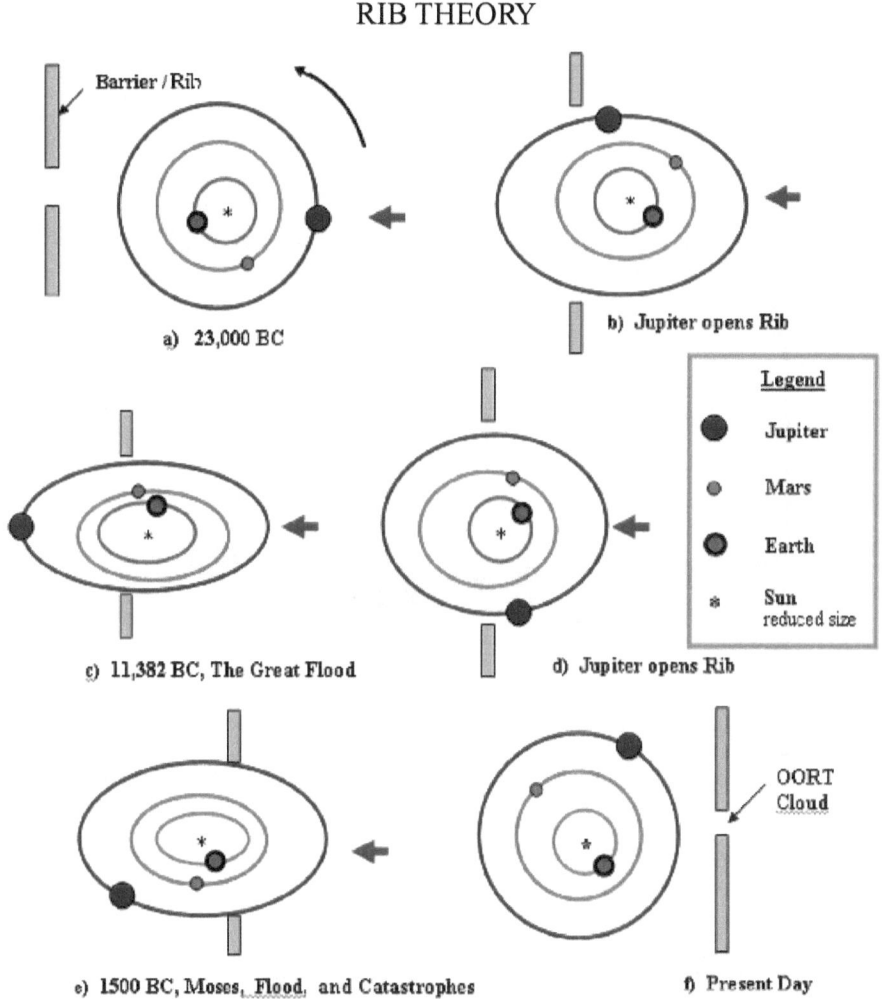

a) 23,000 BC

b) Jupiter opens Rib

c) 11,382 BC, The Great Flood

d) Jupiter opens Rib

Legend

- Jupiter
- Mars
- Earth
- Sun reduced size

e) 1500 BC, Moses, Flood, and Catastrophes

f) Present Day

Rib Theory is based upon:

1. The Time Line, decoded from the Bible. 2. A barrier that surrounds each spiral sector in our Galaxy.

3   Jupiter having the ability to open the rib with more influence than Saturn, Uranus, Neptune or Pluto.

Since the barrier/rib doesn't lend itself to the passage of solar systems, the aperture opening tries to close when Jupiter isn't near. When the Solar System was passing through the barrier and the aperture opening was at its minimum, the following happened in frame 7c:

a)   Mars released oceans of water to the gravitational pull of Earth when their orbits were elliptical.

b)   Some water settled near the Oort cloud where it resides today.

The Earth and Mars moved closer to the Sun by 32 (26%) and 108 (43%) million miles, respectively. See fig 5

Figure 7

Chapter 7

SUPPLEMENTAL READING

There are three phenomena on Earth's surface surrounded in a mystique with explanations unrelated to a real cause and effect. They are documented with illustrations in the chapters titled Earth's Phenomena. Statistical information and ancient beliefs are noted for each phenomenon. Photographs are taken from a royalty free Web site. Pointers or URL addresses are used to indicate where the reader can see additional photos in accordance with copyright laws.

The first and second topics concern Meteor Crater, Arizona, and Ayers Rock, Australia. The previous chapters developed a hypothesis and an axiom for the elliptical orbital paths of Earth and Mars. The author Velikovsky writes about the near collision between Earth and Mars and again between Earth and Venus. With a time line obtained from the Bible, the first near collision between Earth and Mars occurred around 11,000 BC—a period of great flooding. In his book titled *The Making of Mankind*, Richard E. Leakey reports on page 37:

> Around 10,000 years ago the water level was 100 meters (330
> feet) higher than it is at present. Why it dropped so precipitously
> is something of a mystery, but several other East Africa lakes
> shrank dramatically at about the same time.

On pages 206 and 207, he discusses how the people in the Near East engaged in intensive food production even though vegetation was sparse ten thousand years ago. Then as the climate warmed, not instantaneously but steadily over the centuries, each generation noted the advancement of wild wheat and barley that spread across the fertile crescent. The difference between 11,000 BC and ten thousand years ago is three thousand years. Either the Bible's time line is slightly incorrect or Mr. Leakey's estimates were rounded off. In either event, he describes a time period when the water level was much higher based upon geological markings and his personal observations in Africa. It is logical to conclude that vegetation was sparse for a long time after a major flood, and that an error difference of three thousand years is slight. Consequently, both statements compliment and support each other.

Meteor Crater Photograph Discussion

Velikovsky writes about floods at the time of Moses, but both Moses and Leakey report devastating flood several thousand years before man learned how to write. Ancient mythology talks about Zeus and his thunderbolt. When we compile all this information, we can speculate that Mars contributed its oceans, its surface boulders, and its voltage potential in the form of a massive lightning bolt. The lightning bolt had to be several feet in diameter when it came down vaporizing Earth's surface and leaving a crater with very little residue. The original photo of figure 8.2 in color has a rainbow of colors in an area one-third the distance below the surface. This coloring of the soil can't be caused by the heat impact of a meteor but rather a lighting bolt. The soil area outside the crater is reddish due to an electrical discharge. These phenomenon correlate with Grover Gilbert's discoveries. In 1891, Grove Karl Gilbert was the chief geologist for the U.S. Geological Survey. He found very little residue and had trouble justifying a meteor as the cause. Grover Gilbert's comments are in chapter 8.

Over twenty-four craters on Earth can be viewed at *www.lpi.usra.edu*, and none have the same characteristics as Meteor Crater, Arizona. They are too old, too big, or too typical in shape with a high outer ridge. Meteor Crater is 0.74 miles across, 656 feet deep, and twenty thousand years old. There isn't any evidence of material residue at the center or the ridge. If correct mathematical attributes were chosen for the crater, it would show that Meteor Crater is different than all others. The attribute would have to be a combination of ratios between the crater's diameter, depth, and ridge height.

The most likely cause for the creation of Meteor Crater is that a voltage potential between Earth and Mars manifested itself over thirteen thousand years ago at the time of the great flood in the form of a massive lighting bolt. Electricity at small voltages can vaporize the metal tip of a screwdriver. Lighting strikes can demolish trees, cactus, and homes. So why can't two spheres huge in mass develop enough static electricity to vaporize the entire volume of the crater? The difference between the Bible's time line and the creation of the crater is seven thousand years. The same justifications that were used to explain error differences between the time line and scientific estimates can be used again.

Ayers Rock Photograph Discussion

Ayers Rock is basically Martian soil. It has the right color, and it sits on flat land, void of any other mountains. Scientist suspect that it failed to crash on Earth because it arrived at a tangential trajectory, started to leave Earth's proximity, and then reversed course, settling down to a soft landing. Figure 8.4 in chapter 8 has a mysterious hollow or indentation. The four pockmarks shown in figure 8.5 resemble the crater marks on the surface of Mars that were distorted when Ayers Rock landed on Earth. (To see additional pockmarks from another view, the reader is asked to use the pointer in chapter 8 and go to the Web site as indicated.) Australian legend states that a lizard scooped out the surface of the rock leaving a series of bowl-shaped hollows. In 2300 BC, Mars came dangerously close to Earth again. It dumped some more water and rocks but not as much as it did thirteen thousand years ago. Venus was the major cause of upheavals during the time of Moses. It shifted the tectonic plates under Egypt causing volcanic eruptions at Santorini that contributed to the plagues described in Exodus. If the tectonic plates shifted as scientist point out today, then our Earth did come dangerously close to colliding with Venus in 1500 BC.

Nazca Lines Photograph Discussion

Extraterrestrial visitors that came here to witness mankind's behavior created the Nazca Lines. The pictographs were a product of the local natives. The Internet is filled with a multitude of pictographs, created by natives and less filled with the geometry of the lines created by visitors. (The reader is asked to use the pointer in chapter 8 to see a map of Ingenio Valley.) The photograph of Ingenio Valley illustrates seven out of ten runways in trapezoid shape, decreasing in width from east to west. An attempt is made by the author to illustrate what is meant by "trapezoidal" without duplicating the map.

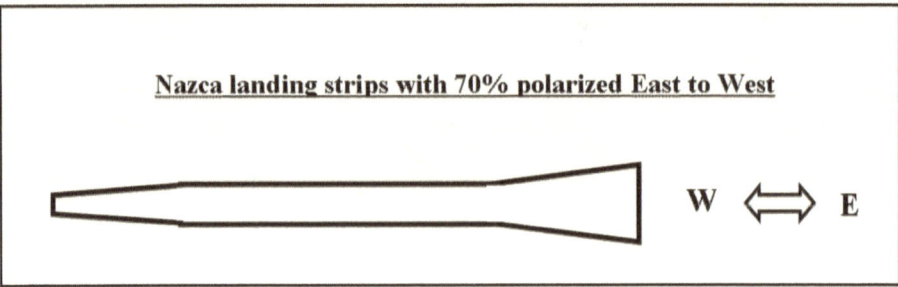

The orientation of these runways is a confirmation of forced landings as discussed in earlier chapters. It reinforces the concept for the requirement of the spacecraft to shift from imaginary time to real time and then land.

Extra Money to Invest

If I was Uncle Sam and had billions of dollars to invest, I would spend it on the following projects:

Phase 1:
I would authorize 3 million dollars for a company like the Naval Laboratory in China Lake, California, to distribute 1 million to three different companies to develop the time shifter. I would stipulate the following conditions in a contract:

1. Each company should know that they are in competition with two other companies whose identity is being kept a secret.
2. Each company is required to have one physicist and one aeronautical engineer within the group of four highly professional people.
3. Each company is given a time frame of six to twelve months to demonstrate that a time shifter can be made to disappear.
4. All experiments are to be performed at late afternoon or at night to circumvent the probability of the plastic prototype disappearing and reappearing in a bulkhead or inside a person.

Phase 2:
With phase one as a success, I would authorize a group of physicist to study the neutrino. One of the best companies is Fermi Labs in Batavia, Illinois. I would stipulate that they determine the following:

1. Neutrinos are abundantly available with speeds up to one hundred times the speed of light, and that they have a Gaussian distribution.

2. The speed of the neutrino can be calculated as the vector sum of its parent, the neutron colliding with other particles at light speed, its forward velocity due to the explosion on the sun, and its phase velocity.

3. Neutrinos can be stopped with special material that's very thin.

4. More neutrinos can be stopped if the material has a forward speed of 0.1% the speed of light.

5. To inform others of their findings, the physicist need to contact an illustrator who can produce an animated film of the neutrino and how its speed is achieved. The film is required for additional funding.

6. The allotted time is two years with 10 million dollars plus equipment expenditures.

Phase 1a, on going:

1. Require that the first group install cameras inside the phase shifter to document characteristics of the GIT curve.

2. Allotted time is 6 months to collect data.

Phase 3:

With the completion of these two feasibility studies and positive results, I would authorize the construction of a spacecraft and engine like the one envisioned in figure 3. I would contract with a manufacturing company to do the following:

1. Review the IBM patent that was mentioned in this book.

2. Modify the patent to become a part of an electromagnetic field.

3. Consult with more physicists to develop an unmanned spacecraft with its own initial propulsion to lift it off the ground or with a plan to drop it from a plane.

4. Collect massive amounts of data and be prepared to launch the first unmanned spacecraft in outer space.

5. Build a second spacecraft that will accommodate astronauts.

6. The allotted time is five years with 500 million dollars plus equipment expenditures.

Phase 3a on going:

I would start training astronauts to perform helicopter landing at night with simulators that duplicate the effects of imaginary time.

1. They have to learn how to correlate a parsec of time with the reduced gravity pull as per the GIT curve coupled with *secondary* light as per chapter 1.

2. After the simulated landing of a spacecraft is mastered, they have to prepare for the actual launch.

The first launch into outer space consists of time shifting in and out of imaginary time and catapulting the spacecraft with astronaut through Earth. This could become a feat more dangerous than the trip that was made to the moon back in 1969. The goal is to leave at night on Earth's wake side, pierce through to Earth's morning side, sail back to the night side, and do a helicopter landing. This sounds very difficult, but if the two feasibility studies are accurate, then the safety factor increases. Feasibility becomes a reality when the properties of the time shifter are understood and a simulator provides proper training.

Phase 4 Diplomatic Studies:

In regard to the hypothesis that were mentioned in chapters 5 and 6, I would authorize expenditures to do the following:

1. Send out emissaries to consult with scientist in Japan and Italy and determine if all neutrinos radiate directly out of the sun or if some turn and travel back into the direction of the black hole. This goal can be achieved by counting the number of neutrinos that arrive every day of the year and coordinate this data with the information mentioned in this book.

2. Hire geologist to determine the following:

   - Could Ayers Rock in Australia be a part of Mars?
   - Was Meteor Crater in Arizona caused by huge electrical discharge?
   - Could the lines at Nazca have been created by a force field from a spacecraft or a sonic boom?

3. Hire astronomers to determine the existence of sector time based upon Rib theory that leads to the following:

   - The location of the Oort clouds where the sun and nine planets squeezed through an opening in the barrier rib
   - The elliptical orbits of the nine planets and Xena with the relationship of their orbital plane with respect to the ecliptic plane of the solar system
   - The possibility of our Earth continuing decay toward the sun and/or its orbital path becoming more circular and less elliptical

In conclusion, this is a monumental task that could take eight years to complete as illustrated in the PERT chart. The motivation is as great as all the government contracts that NASA and other agencies have justified when they were awarded their present-day contracts. If a better method than rocket power for traveling through outer space is possible, then why not start with a small expenditure like phase one and prove the existence of imaginary time?

PERT Chart

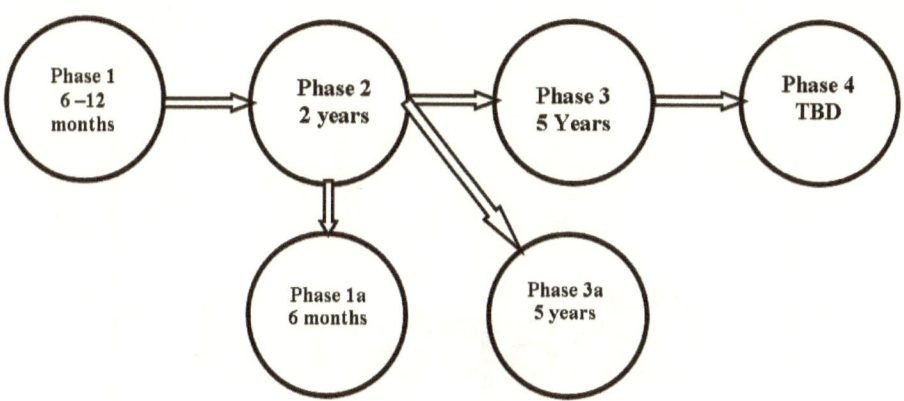

As a final note, Einstein said in so many words that it's "all relative." Because it's all relative, mankind always has a choice in situations like whether the sun goes around Earth or whether Earth goes around the sun, for example. Man never weighs his choices and never gets it right more than 50% of the time. I can only hope and pray that my solutions are the correct ones because mankind will have a better chance for understanding the universe and our galaxy. To the least extent, it is hoped that mankind begins to see the thin thread of continuity that connects all media and news articles to the development of a new frontier—the same goal toward a frontier that was stated in the beginning of this book.

Chapter 8

EARTH'S PHENOMENA

Meteor Crater, Arizona

Figure 8.1
Credit: Maunger at Dreamstime.com

Meteor Crater, Arizona

Figure 8.2

Credit: JLVDream at Dreamstime.com

Statistics:

- 1.2 kilometers (0.74 miles) in diameter
- 200 meters (656 feet) deep
- Formed 49,000 years ago as per first source
- Formed 20,000 to 50,000 years ago as per second source

Information:

Text information taken from the Internet at *http://www.barringercrater.com/science*.

"In 1891 Grove Karl Gilbert, then chief geologist for the U.S. Geological Survey, decided to test two conflicting hypotheses about the crater. The first was that the crater was created by the impact of a giant meteorite; the second, that it was the result of an explosion of superheated steam, caused by volcanic activity far below the surface."

"If an iron meteorite had created the crater, Gilbert assumed that it would have had to be nearly as big as the crater itself. So what predictions could he test?"

"First, the meteorite should be taking up a lot of space in the hollow of the crater. The volume of the hollow would therefore be less than the volume of the ejected material in the crater rim. Second, the presence of a large mass of buried iron should affect the behavior of magnets and compass needles. **Neither prediction was confirmed.** Gilbert concluded that a steam explosion was the only surviving hypothesis, in spite of the fact that no volcanic rocks had ever been found in the area. The meteorites around the crater were simply a coincidence."

EARTH'S PHENOMENA

Ayers Rock (Mount Uluru) Australia

Figure 8.3

Credit: Hdsidesign at Dreamstime.com

Ayers Rock (Mount Uluru) Australia

Figure 8.4
Credit: Andb068 at Dreamstime.com

Ayers Rock (Mount Uluru) Australia

Figure 8.5

Credit: Slovegrove at Dreamstime.com

Pointer: *http://www.crystalinks.com/ayers1.gif*
- Illustrates four craters within Ayers Rock

Statistics:
- Coarse-grained sandstone rich in feldspar
- Height: 318 meters (986 feet)
- Circumference: 8 kilometers (5 miles) Translates to: 4,200 feet across
- Depth: Extends 2.5 kilometers (1.5 miles) into the ground

Information:

Text information taken from the Internet at *http://www.crystalinks. com/ayersrock.html*.

For Australian Legend

(Log onto Internet, copy pointer address, and paste address to see photograph.)

"In the creation period, Tatji, the small Red Lizard, who lived on the mulgi flats, came to Uluru. He threw his kali, a curved throwing stick, and it became embedded in the surface. He used his hands to scoop it out in his efforts to retrieve his kali, leaving a series of bowl-shaped hollows."

EARTH'S PHENOMENA

Nazca Lines and Pictographs

Figure 8.6

Credit: Tacna at Dreamstime.com

Nazca Lines and Pictographs

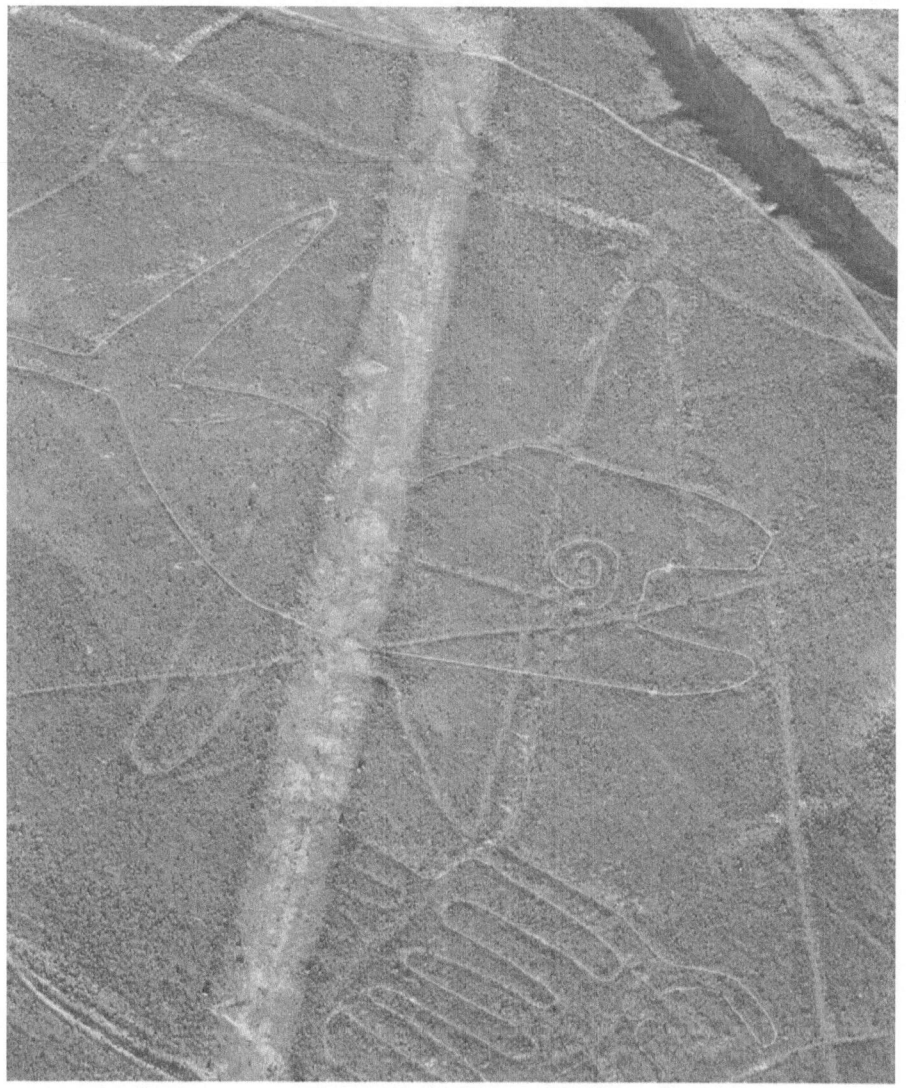

Figure 8.7

Credit: Jarnog at Dreamstime.com

Pointer: *www.nazcamystery.com/images/nazca_22.gif* or
http://theflashpackers.blogspot.com/2007/10/nazca-lines.html or
http://www.ladatco.com/NAZ-map.jpg

- Three different Web sites illustrate a map of the Ingenio Valley or the Nazca Valley.
- Seven out of ten runways aligned in an east to west direction.

Pointer: *http://www.orquidea.net/nazca_lines_small.jpg* or
http://www.mars-earth.com/nazca_pad.jpg

- Illustrates a trapezoidal runway.

Information:

Text information taken from the Internet at UNESCO

Declared "**Archaeological World Heritage Site**" by UNESCO (1994)

The Nazca Lines are located in the arid Peruvian coastal plain, some 400 km south of Lima, the geoglyphs of Nazca and the pampas of Jumana cover about 450 sq. km.

The Nazca Lines, which were scratched on the surface of the ground between 500 BC and AD 500, are among archaeology's greatest enigmas because of their quantity, nature, size, and continuity.

The geoglyphs depict living creatures, stylized plants and imaginary beings, as well as geometric figures several kilometers long. They are believed to have had ritual astronomical functions.

GLOSSARY OF TERMS
AND CONCEPTS

Atomic Number:

An experimentally determined number characteristic of a chemical element that represents the number of protons in the nucleus, which in a neutral atom equals the number of electrons.

Atomic Unit or Atomic Weight:

The average atomic mass of an element compared to 1/12 the mass of carbon 12. The hydrogen atom consists of a proton and an electron has an atomic number of one and an atomic weight of 1.0079.

Axiom:

A statement accepted as true as the basis for an argument or inference.

Biological Clock:

An inherent timing mechanism that is inferred to exist in some living systems in order to explain various cyclical behaviors and physiological processes.

dBc:

Symbol for "decibels below carrier." A power rating used to describe the purity of a sinusoidal waveform or pattern by measuring the distance between the fundamental carrier and the harmonics.

Earth's Wake:

A location in space that Earth has recently traveled through.

Elysian Fields:

Defined by Edith Hamilton in her book *Mythology* as a place where everything is delightful, with soft green meadows, lovely groves, and sunlight that glowed softly purple, an abode for dead heroes and great men who made men remember them by helping others.

GIT Curve:

An abbreviation for "gravity versus imaginary time." A mathematical expression that relates gravity to imaginary time where the units are percent change and parsecs, respectively.

Imaginary Time:

Mathematically equal to the symbol $\pm\sqrt{-1}*T$ or $\pm jT$ where $j = \sqrt{-1}$ or $= (-1)^{1/2}$. Conceptually equal to a frontier, a location, a time warp in another dimension.

Forced Landing:

A condition associated with a spacecraft that needs to land when Earth appears to have a relative speed of one thousand miles per hour.

Helicopter Landing:

A condition associated with a spacecraft that lands on Earth at nighttime by entering Earth's wake.

Hypothesis:

A tentative assumption made in order to draw out and test its logical or empirical consequences.

Mass:

Mass is the weight of an object. In the English system mass is in pounds. In the metric system mass is in grams or kilograms.

Parsec:

A measurement of imaginary time that could be equal to an hour, a minute, or a second, henceforth the name "part of a second" or parsec.

Propagation Time:

A condition that varies with material whereby the speed of a radio wave slows down from the speed of light in free space to 60% of its original value as it passes though a material.

Sector Time:

A time frame within a sector of the galaxy where the transition of time is faster or slower with respect to another sector.

Speed of Light:

Equal to 186,000 miles per second or 3×10^8 meters per second.

Standard Deviation:

Standard Deviation (or Stdev) is the square root of variance. Variance is the arithmetic mean of the squared deviations of individual items about their mean.

$Stdev = [\ \Sigma(X - u)^{\wedge 2} / N \ ]^{\wedge}.5$ where u is the mean, X is the deviation, and N is the number individuals or events.

- The mean ± 1 Stdev = 68.3% of the distribution.
- The mean ± 2 Stdev = 95.4% of the distribution.
- The mean ± 3 Stdev = 99.7% of the distribution

Rib:

A barrier of the galaxy that separates the sectors. Proposed as an invisible stream of neutrinos at the sector's edge causing the galaxy to have skeletal rigidity.

Rib Theory:

A hypothesis with limited proof about the transition of our solar system migrating across sectors.

Theory:

An ideal or hypothetical set of facts.

Theorem:

An idea accepted or proposed as a demonstrable truth.

Time Shifter:

A device that is capable of moving through imaginary time and thereby reducing the effects of gravitational pull. See GIT curve.

Warp Speed:

The speed of an object relative to the speed of light with the assumption of linearity. For example, Warp 1 would equal the speed of light, Warp 2 would equal twice the speed of light, and so on and so forth.

BIBLIOGRAPHY

Books

Hamilton, Edith, *Mythology,*
  Little Brown and Company, original 1942; with paper back in 1998

Hancock, Graham, *Finger Prints of the Gods,*
  THREE RIVERS PRESS, Crown Publishers, Inc. 1995

Hawking, Stephen, *A Briefer History of Time,*
  Bantam Books, Random House, Inc., 2005

Jordan, Edward C., *Electromagnetic Waves and Radiating Systems,*
  Prentice-Hall, Inc. 1950

Leakey, Richard E., *The Making of Mankind,*
  Rainbird Publishing Group Limited, Division of E.P. Dutton,
  Elsevier-Dutton Publishing Co., Inc. 1981

McGraw-Hill, *Electrical and Electronic Engineering Series, by
  Hayt, William H. Jr., Engineering Electromagnetic*
  McGraw-Hill Book Company, Inc. 1958

Von Daniken, Erich, *Chariots of the Gods,*
    Souvenir Press Ltd., Great Britian,1969
    Ryerson Press, Canada, 1969

G. P. Putnam's Sons, New York,1969
    Velikovsky, Immanuel, *Worlds In Collision,*
    DoubleDay & Company, Inc. 1950

Quoted Magazines

Berman, Bob. "Confused About Your Direction." *Discovery*
    (September 2006): p. 33

Powell, Corey S. "What's at the Edge of the Solar System." *Discovery*
    (May 2006): p. 44

Weinstock, Maia. "X-ray Eye on Dark Matter." *Discovery*
    (December 2001): p. 16

Wright, Karen. "Ripples in Spacs." *Discovery* (May 2001): p. 18

Svitil, Kathy. "The Great Buried Floods of Masr." *Discovery*
    (June 2000): p. 28

Kunzig, Robert. "The Unbearable, Unstoppable Neutrinop." *Discovery*
    (August 2001): p. 34

DiChristina Mariette. "Space at Warp Speed." *Popular Science*
    (May 2001): p. 48

Other Related Magazines

Lemonick, Michael D. "Star Seeker." *Discovery* (November 2001)

Kunzig, Robert. "Trapping Light." *Discovery* (April 2001)

Other Related Books

Charles T. Clark/Lawrence L. Schkade*Statistical Analysis For Administrative Decisions,* South-Western Publishing Co., 1983 page 35, Chapter 2 and Appendix H

INDEX